AF646307

Encyclopaedia of
Nanotechnology

Encyclopaedia of
Nanotechnology

Vol. 7 : Diagnostic and Therapeutics

Dr. Rita Khare

AGROTECH PRESS

JAIPUR ▪ NEW DELHI

AGROTECH PRESS

Regd. Off. "Bony Residency", Gate No. 2
Opp. Tilak Public School, Vishweswariya Nagar,
Gopalpura Road, Jaipur - 302018, India
Ph: 0141 - 2761280, Telefax: 2761381
E-mail: agrotechpress@gmail.com

Delhi Office: 102, 1st Floor, Satyam House,
4327/3, Ansari Road, Darya Ganj,
New Delhi - 110002, India
Ph: 011-45652440
visit us: www.agrotechpress.com

Distributed by :
KEYA INTERNATIONAL PRESS LLP
E-mail: keyainternationalpressllp@gmail.com

Published in India, 2024

© All rights reserved

ISBN : 9789384568566 (10 Vol. Set)

All Rights Reserved. Neither this publication nor any part thereof may be reproduced, stored in a retrieval system, or transmitted, in any form or by any means, electronic, mechanical, photocopying, recording, scanning or otherwise, without the prior written permission of the copyright owner.

Typeset by Agrotech Press

Printed and bound in India by Replika Press Pvt. Ltd.

Contents

Preface

Nanotechnology has been broadly defined as the one for not only the creation of functional materials and devices as well as systems through control of matter at the scale of 1-100 nm, but also the exploitation of novel properties and phenomena at the same scale. It offers an extraordinary, paradigm-changing opportunity to make significant advances in diagnosis and treatment. In the last several decades, nanotechnology has been studied and developed primarily for use in novel drug-delivery systems. A recent explosion in engineering and technology has led to the development of many new nanoscale platforms, including quantum dots, nanoshells, gold nanoparticles, paramagnetic nanoparticles, and carbon nanotubes, and improvements in traditional, lipid-based nanoscale platforms. The emerging implications of these platforms for advances in diagnostics and therapeutics form the basis of this volume. A widespread understanding of these new technologies is important, because they currently are being integrated into the clinical practice of oncology. Nanomaterials are increasingly used in diagnostic, imaging, and targeted drug delivery applications.

This text describes the state of the art in nanoparticle and nanodevice applications for medical diagnosis and disease treatment. Nanodevices, such as cantilevers, have been integrated into high-sensitivity disease marker diagnostic detectors and devices, are stable over long periods of time, and display reliable performance properties. Nanotechnology strategies have been applied to therapeutic purposes as well. For example, nanoparticle-based delivery systems have been developed to protect drugs from degradation, thereby reducing the required dose and dose frequency, improving patient comfort and convenience during treatment, and reducing treatment expenses. The main objectives for integrating nanotechnologies into diagnostic and therapeutic applications in the context of various diseases are reviewed.

Innovation in the field of nanotechnology is required. The field is, therefore, characterized by the continuous growth and evolution in response to these barriers to product development. As a result of these efforts, nanowires, nanocantilevers, quantum dots, nanoshells, dendrimers, liposomes, nanopyramids, and nanogels are used in diagnostic or therapeutic applications. Extensive types of nanoparticles composed of different materials, shapes, and sizes, and with various chemical and surface properties, have already been engineered. These developments offer exciting opportunities for the development of personalized therapy, in which the molecular profiles of an individual's genetic and protein biomarkers may be used to diagnose and treat the patient's cancer.

Nanotechnology promises to facilitate the development of personalized medicine, in which patient therapy is tailored by the patient's individual genetic and disease profile. This book provides an overview of nanotechnology applications in molecular diagnostics and drug delivery. The coupling of nanotechnology strategies and telecommunication will improve the precision of diagnostics and therapeutics and promise to impact the care and management of diseases such as intestinal diseases and cancer. In the near future, medical approach and diagnostics will certainly dramatically change. Nanoscale sensors and devices may provide a personal and cost-effective continuous medical monitoring of patient's health. Nanoscale sensors and devices may also support an enhanced prediction in early stage of diseases such as cancer or inflammations. Further future perspectives include technology miniaturization which will allow association of nanotools with nanodevices; this will allow novel measurements such as an optical transducer for light; an electro/chemical transducer for electrical properties; a magnetic transducer for changes to local magnetic fields; and a mechanical transducer, to detect changes in motion.

The development of safe and reliable nanomaterials is at the foundation of emerging nanomedicine-based therapies and imaging techniques. A variety of different nanoparticle platforms have been explored as potential delivery vehicles: polymers, liposomes, micelles, emulsions, metal, metal oxide, dendrimers, fullerenes, quantum dots, and carbon nanotubes. Not only have the types of nanomaterials in use broadened widely, but researchers have discerned how to functionalize nanoparticles, characterize complex multifunctional conjugates, and understand effects on their biodistribution and toxicity. Overall, further progress in the field is expected to move along two parallel tracks. First one will be associated with on-going translation to the clinical environment; while the second with the development of new tools and techniques in research arena, where continuous innovation in the area of nanomaterials will lead to more sophisticated and multifunctional nanoparticle and nano-devices designs. This text will encompass all aspects of research related to the field of applied nanotechnology in medicine. The field of applied nanotechnology in medicine is growing fast. To cover the broad aspects of nanotechnology for medical applications and their eco-toxicological effects, there is a need for a book that also aims to improve communication between basic and clinical science. The book discusses advances in drug discovery, drug delivery, and molecular diagnosis and imaging linked to the application of nanotechnology in medicine.

Dr. Rita Khare

1

Introduction

Nanomedicine is the medical application of nanotechnology. Nanomedicine ranges from the medical applications of nanomaterials, to nanoelectronic biosensors, and even possible future applications of molecular nanotechnology. Current problems for nanomedicine involve understanding the issues related to toxicity and environmental impact of nanoscale materials (materials whose structure is on the scale of nanometres, i.e. billionths of a metre). Nanomedicine research is receiving funding from the US National Institutes of Health. Of note is the funding in 2005 of a five-year plan to set up four nanomedicine centres. In April 2006, the journal Nature Materials estimated that 130 nanotech-based drugs and delivery systems were being developed worldwide.

The biological and medical research communities have exploited the unique properties of nanomaterials for various applications (e.g., contrast agents for cell imaging and therapeutics for treating cancer). Terms such as *biomedical nanotechnology*, *nanobiotechnology*, and *nanomedicine* are used to describe this hybrid field. Functionalities can be added to nanomaterials by interfacing them with biological molecules or structures. The size of nanomaterials is similar to that of most biological molecules and structures; therefore, nanomaterials can be useful for both in vivo and in vitro biomedical research and applications. Thus far, the integration of nanomaterials with biology has led to the development of diagnostic devices, contrast agents, analytical tools, physical therapy applications, and drug delivery vehicles.

Nanomedicine seeks to deliver a valuable set of research tools and clinically useful devices in the near future. The National Nanotechnology Initiative expects new commercial applications in the pharmaceutical industry that may include advanced drug delivery systems, new therapies, and in vivo imaging. Neuro-electronic interfaces and other nanoelectronics-based sensors are another active goal of research. Further down the line, the speculative field of molecular nanotechnology believes that cell repair machines could revolutionize medicine and the medical field.

Nanomedicine is a large industry, with nanomedicine sales reaching $6.8 billion in 2004, and with over 200 companies and 38 products worldwide, a minimum of $3.8 billion in nanotechnology R&D is being invested every year. As the nanomedicine industry continues to grow, it is expected to have a significant impact on the economy.

Medical Use of Nanomaterials

Two forms of nanomedicine that have already been tested in mice and are awaiting human trials that will be using gold nanoshells to help diagnose and treat cancer, and using liposomes as vaccine adjuvants and as vehicles for drug transport. Similarly, drug detoxification is also another application for nanomedicine which has shown promising results in rats. A benefit of using nanoscale for medical technologies is that smaller devices are less invasive and can possibly be implanted inside the body, plus biochemical reaction times are much shorter. These devices are faster and more sensitive than typical drug delivery. Advances in Lipid nanotechnology was also instrumental in engineering medical nanodevices and novel drug delivery systems as well as in developing sensing applications.

Drug Delivery

Nanotechnology has provided the possibility of delivering drugs to specific cells using nanoparticles. The overall drug consumption and side-effects may be lowered significantly by depositing the active agent in the morbid region only and in no higher dose than needed. This highly selective approach would reduce costs and human suffering. An example can be found in dendrimers and nanoporous materials.

Another example is to use block co-polymers, which form micelles for drug encapsulation. They could hold small drug molecules transporting them to the desired location. Another vision is based on small electromechanical systems; nanoelectromechanical systems are being investigated for the active release of drugs. Some potentially important applications include cancer treatment with iron nanoparticles or gold shells. Targeted drug delivery is intended to reduce the side effects of drugs with concomitant decreases in consumption and treatment expenses. The increased efficiency of delivery results in overall societal benefit by reducing the amount of drug needed in an equipotent preparation of said therapy, and thus reduced cost to the consumer.

Nanomedical approaches to drug delivery centre on developing nanoscale particles or molecules to improve drug bioavailability. Bioavailability refers to the presence of drug molecules where they are needed in the body and where they will do the most good. Drug delivery focuses on maximizing bioavailability both at specific places in the body and over a period of time. This can potentially be achieved by molecular targeting by nanoengineered devices. It is all about targeting the molecules and delivering drugs with cell precision. More than $65 billion are wasted each year due to poor bioavailability. *In vivo* imaging is another area where tools and devices are being developed. Using nanoparticle contrast agents, images such as ultrasound and MRI have a favourable distribution and improved contrast. The new methods of nanoengineered materials that are being developed might be effective in treating illnesses and diseases such as cancer. What nanoscientists will be able to achieve in the future is beyond current imagination. This might be accomplished by self assembled biocompatible nanodevices that will detect, evaluate, treat and report to the clinical doctor automatically.

Drug delivery systems, lipid- or polymer-based nanoparticles, can be designed to improve the pharmacological and therapeutic properties of drugs. The strength of drug delivery systems is their ability to alter the pharmacokinetics and biodistribution of the drug. However, the pharmacokinetics and pharmacodynamics of nanomedicine is highly variable among different patients. When designed to avoid the body's defence mechanisms, nanoparticles have beneficial properties that can be used to improve drug delivery. Where larger particles would have been cleared from the body, cells take up these nanoparticles because of their size. Complex drug delivery mechanisms are being developed, including the ability to get drugs through cell membranes and into cell cytoplasm. Efficiency is important because many diseases depend upon processes within the cell and can only be impeded by drugs that make their way into the cell. Triggered response is one way for drug molecules to be used more efficiently. Drugs are placed in the body and only activate on encountering a particular signal. For example, a drug with poor solubility will be replaced by a drug delivery system where both hydrophilic and hydrophobic environments exist, improving the solubility. Also, a drug may cause tissue damage, but with drug delivery, regulated drug release can eliminate the problem. If a drug is cleared too quickly from the body, this could force a patient to use high doses, but with drug delivery systems clearance can be reduced by altering the pharmacokinetics of the drug. Poor biodistribution is a problem that can affect normal tissues through widespread distribution, but the particulates from drug delivery systems lower the volume of distribution and reduce the effect on non-target tissue. Potential nanodrugs will work by very specific and well-understood mechanisms; one of the major impacts of nanotechnology and nanoscience will be in leading development of completely new drugs with more useful behaviour and less side effects. Polymeric nano-particles are a competing technology to lipidic (based mainly on Phospholipids) nano-particles. There is an additional risk of toxicity associated with polymers not widely studied or understood.

This toxicity could include (but not limited to) hepatotoxicity, nephrotoxicity etc. and can have long term impacts not easily evaluated in short term in-vivo clinical trials either in animals or humans. Since the degradation of polymers to either their monomers or other degradation products in the body cannot be accurately predicted (unclear metabolic pathway), this is a real risk, specially in medicines intended for long term patient use. Even if the toxicity is ignored, there is additional hepatic load of metabolism, again an area of concern for long term medication. The major advantages of polymers is stability, lower cost and predictable characterisation, so the formulation chemists prefer this. However, in the patient's body this very stability (slow degradation) is a negative factor. Phospholipids on the other hand are membrane lipids (already present in the body and surrounding each cell), have a GRAS (Generally Recognised As Safe) status from FDA and are derived from natural sources without any complex chemistry involved. They are not metabolised but rather absorbed by the body and the degradation products are themselves nutrients (fats or micronutrients). It is greatly observed that nanoparticles are promising tools for the advancement of drug delivery, medical imaging, and as diagnostic sensors. However, the biodistribution of these nanoparticles is still imperfect due to the complex host's reactions to nano- and microsized materials and the difficulty in targeting specific organs in the body.

Nevertheless, a lot of work is still ongoing to optimize and better understand the potential and limitations of nanoparticulate systems. For example, current research in the excretory systems of mice shows the ability of gold composites to selectively target certain organs based on their size and charge. These composites are encapsulated by a dendrimer and assigned a specific charge and size. Positively-charged gold nanoparticles were found to enter the kidneys while negatively-charged gold nanoparticles remained in the liver and spleen. It is suggested that the positive surface charge of the nanoparticle decreases the rate of opsonization of nanoparticles in the liver, thus affecting the excretory pathway. Even at a relatively small size of 5 nm, though, these particles can become compartmentalized in the peripheral tissues, and will therefore accumulate in the body over time. While advancement of research proves that targeting and distribution can be augmented by nanoparticles, the dangers of nanotoxicity become an important next step in further understanding of their medical uses.

Nanoparticles are also used to circumvent multidrug resistance (MDR) mechanisms. Mechanisms of MDR include decreased uptake of drugs, reduced intracellular drug concentration by activation of the efflux transporters, modifications in cellular pathways by altering cell cycle checkpoints, increased metabolism of drugs, induced emergency response genes to impair apoptotic pathways and altered DNA repair mechanisms.

Applications and Reported Research Studies

- Abraxane, approved by the U.S. Food and Drug Administration (FDA) to treat breast cancer, non-small- cell lung cancer (NSCLC) and pancreatic cancer, is the nanoparticle albumin bound paclitaxel.
- Doxil was originally approved by the FDA for the use on HIV-related Kaposi's sarcoma. It is now being used to also treat ovarian cancer and multiple myeloma. The drug is encased in liposomes, which helps to extend the life of the drug that is being distributed. Liposomes are self-assembling, spherical, closed colloidal structures that are composed of lipid bilayers that surround an aqueous space. The liposomes also help to increase the functionality and it helps to decrease the damage that the drug does to the heart muscles specifically.
- In a mice study, scientists from Rice University and University of Texas MD Anderson Cancer Centre reported enhanced effectiveness and reduced toxicity of an existing treatment for head and neck cancer when using the nanoparticles to deliver the drug. The hydrophilic carbonic clusters functionalized with polyethylene glycol or PEG-HCC are mixed with the chemotherapeutic drug paclitaxel (Taxol) and the epidermal growth factor receptor (EGFR) targeted Cetuximab and injected intravenously. They found the tumors were killed more effectively with radiation and the healthy tissue suffered less toxicity than without the nanotechnology drug delivery. The standard treatment contains Cremophor EL which allows the hydrophobic paclitaxel to be delivered intravenously. Replacing the toxic Cremophor with carbon nanoparticles eliminated its side effect and improved drug targeting which in turn required a lower dose of the toxic paclitaxel.
- Researchers at Case Western Reserve University reported using nanoparticle chain to deliver doxorubicin to breast cancer cells in a mice study. Three magnetic, iron-oxide nanospheres were chemically linked to one doxorubicin-loaded liposome and formed a 100 nm long nanoparticle chain. After the nanochains penetrated the tumor, radiofrequency field was generated that caused the magnetic nanoparticles to vibrate and rupture the liposome, dispersing the drug in its free form throughout the tumor. The result showed that the nano treatment was more effective in halting tumor growth than the standard treatment with doxorubicin. It is also less harmful to healthy cells since only 5% to 10% of the standard dose of doxorubicin were used.

- Nanoparticles made of polyethylene glycol (PEG) carrying payload of antibiotics at its core could swift charge thus allowing them to target bacterial infection more precisely inside the body, a group of MIT researchers reported. The nanoparticles, containing a sub-layer of pH sensitive chains of the amino acid histidine, carry a slightly negative charge when circulating in the blood stream, can evade detection and clearing by the immune system. When they encounter an infection site the particles gain a positive charge provoked by the slightly acidic environment at the infection sites, allowing them to bind to the negatively charged bacterial cell walls and release antibiotics at locally high concentration. This nano delivery system can potentially destroy bacteria even if it has developed resistance to antibiotics because of the targeted high dose and prolonged release of the drug. Although a lot of work is still needed, the researchers believe that it points to a new direction of using nanotechnology to treat infectious disease.
- Using the biomimetic strategy, researchers in the Harvard University Wyss Institute demonstrated in a mouse model that the drug coated nanoparticles can dissolve blood clots by selectively binding to the narrowed regions in the blood vessels – just like the platelets do. Aggregates of biodegradable nanoparticles coated with tissue plasminogen activator (tPA), each about the size of a platelet, were injected intravenously. In the region of vessel narrowing, shear stresses dissociate the aggregates and release the tPA-coated nanoparticles which bind and degrade the blood clots. By précised targeting and concentrating drug at the location of obstruction, the dose used is less than 1/50th of the normal dose. The nanotherapeutics will greatly reduce the severe side effect of bleeding, commonly found in standard treatment of thrombosis.
- The X-shaped RNA nanoparticles capable of carrying four functional modules were created by researchers in the University of Kentucky. These RNA molecules are chemically and thermodynamically stable, able to remain intact in the mouse body for more than 8 hours and to resist degradation by RNase in the blood stream. With its four arms attached with a combination of different active agents, for example, siRNA (for gene silencing), microRNA (for gene expression regulation), aptamer (for targeting) and ribozyme (as catalyst), the X-shaped RNA can achieve therapeutic and diagnostic functions by regulating gene expression and cellular function, and binding to cancer cells with precision, enhanced by its polyvalent nature and synergistic effects by design.
- An early phase clinical trial using the platform of 'Minicell' nanoparticle for drug delivery have been tested on patients with advanced and

untreatable cancer. Built from the membranes of mutant bacteria, the minicells were loaded with paclitaxel and coated with cetuximab, antibodies that bind the epidermal growth factor receptor (EGFR) which is often overexpressed in a number of cancers, as a 'homing' device to the tumor cells. The tumor cells recognise the bacteria from which the minicells have been derived, regard it as invading microorganism and engulf it. Once inside, the payload of anti-cancer drug kills the tumor cells. Measured at 400 nanometres, the minicell is bigger than synthetic particles developed for drug delivery. The researchers indicated that this larger size gives the minicells a better profile in side-effects because the minicells will preferentially leak out of the porous blood vessels around the tumor cells and do not reach the liver, digestive system and skin. This Phase 1 clinical trial demonstrated that this treatment is well tolerated by the patients. As a platform technology, the minicell drug delivery system can be used to treat a number of different cancers with different anti-cancer drugs with the benefit of lower dose and less side-effects.

- Researchers at The Methodist Hospital Research Institute in Houston have created 'LeukoLike Vectors' or LLVs. LLVs are the drug-carrying silicon nanoparticles coated with the lipoprotein removed from the membranes of the white blood cells, leukocytes. The 'cloaked' nanoparticles were found to behave like leukocytes, are able to evade the body's immune system and survive much longer *in vivo* when studied in mice. The scientists expressed that LLVs are able to overcome a major hurdle in nanomedicine delivery by avoiding clearance in blood stream, crossing the biological barriers and localizing at target tissues due to its leukocyte like characters. Synthetic membranes are expected to develop in the future to replace harvest from the white blood cells due to the limited source of leukocyte.
- Inhalable nanocarriers were developed using a nanostructured lipid nanocarrier-based system (NLCS). The lipid nanoparticles contain the drug payload (doxorubicin or paclitaxel); two types of small interfering RNA (siRNA, targeted the BCL-2 mRNA to suppress nonpump cellular resistance, and MRP1 mRNA to suppress pump drug resistance); and the targeting molecules, an analog of luteinizing hormone-releasing hormone (LHRH) which homes in the overexpressed receptors of the cancer cells. The disease model tested was a mouse model of human lung cancer. Results showed enhanced antitumor activity and significant reduction of exposure of healthy organs when compared to the conventional intravenous treatment.
- Self-powered nanoparticles with a payload of osteoporosis drug are reported to be pulled toward the fresh bone crack site, powered by the

ion gradients generated by leaked minerals (ions) from the bone, and deliver highly targeted drugs for healing. Using a model system of human tibia and femur, researchers from Pennsylvania State University and Boston University used nanoparticles of polylactic-coglycolic acid (widely used in medical devices) carrying sodium alendronate (an osteoporosis drug) for the *in vitro* experiments and found the 'nanotrucks' successfully carry the drug to the fresh crack of the human bone.

- Nanodiamond is diamond particle with a diameter of 2 to 8 nm and surface functional groups to bind various kinds of molecule. Clusters of the nanodiamonds in the range of 50-100 nm in diameter can be manufactured to trap drug inside and slowly release the payload. Experiments by researchers in Northwestern University on mouse models with chemotherapy-resistant liver and breast cancers show that the nanodiamonds with a normally lethal dose of doxorubicin significantly reduce the size of the tumors without toxic side effects expected from the drug if administered alone. The nanodiamond drug complex is found to have a much longer half-life in circulation without lowering the white blood cell count and, once inside the tumor, the nanoparticles are retained much longer. Additional studies using nanodiamonds include treating the triple-negative breast cancer with epirubicin and glioblastoma with doxorubicin. Increase in survival rate with reduced side effects were observed in both the mouse models.
- In regenerative medicine research, silica based nanoparticles are used to deliver molecules to transplanted cells derived from stem cells. Using a mouse model, scientists delivered two synthetic versions of neurotrophic growth factors locally using mesoporous nanoparticles to motor neurons derived from transplanted embryonic stem cells and reported improved long term survival and functional differentiation of the transplant *in vivo.*
- Nanoparticle with a liposomal core and a gel like shell carrying two different types of drug is demonstrated to shrink breast cancer tumor in a mouse model. Researchers in the North Carolina State University designed this nanoparticle, called Gelipo, to deliver both drugs simultaneously and maximize the efficiency of both. The cross-linked hyaluronic acid (HA) shell holds and protects the tumor necrosis factor-related apoptosis-inducing ligand (TRAIL) while the core carries liposomes filled with doxorubicin embedded with peptide. HA preferentially concentrates in hyaluronidase (HAase) rich tumor environment and interacts with receptors on the cancer cell membranes. When the HAase degrades the gel shell of Gelipo, the TRAIL protein is released onto the cell membrane and triggers apoptosis (cell death). The core liposomes are internalized by the tumor cells into endosomes,

the cellular transport system. The peptide breaks down the endosome and releases doxorubicin intracellularly, maximizes the efficacy while minimizes the toxic side effects.

Protein and Peptide Delivery

Protein and peptides exert multiple biological actions in the human body and they have been identified as showing great promise for treatment of various diseases and disorders. These macromolecules are called biopharmaceuticals. Targeted and/or controlled delivery of these biopharmaceuticals using nanomaterials like nanoparticles and Dendrimers is an emerging field called nanobiopharmaceutics, and these products are called nanobiopharmaceuticals.

Applications and Reported Research Studies

- Nanoparticles delivering the myelin antigens were found to induce immune tolerance in a mouse model with relapsing multiple sclerosis. Biodegradable polystyrene microparticles coated with the myelin sheath peptides reset the mouse's immune system and prevent the disease from recurring or reduce the symptoms by halting the attack of the immune system to the protective myelin sheath coating the nerve fibers of the central nervous system. Team of researchers in Northwestern University indicated that this treatment method can potentially be used in other autoimmune diseases.
- Researchers at UCLA developed a nanoparticle system with water-soluble shell encapsulated a protein derived from bird anemia virus, apoptin. Apoptin selectively sends 'self-destruct' signal to tumor cells and trigger the programmed cell death (apoptosis) as it builds up in the nucleus, while leaving healthy cells unharmed. In the mouse model of human breast cancer, the treatment was found to significantly reduce the tumors' growth. This new form of treatment resembles chemotherapy and gene therapy but without the risk of harming healthy cells, a common pitfall of chemotherapy, and no probability of genetic mutation that usually accompanies gene therapy.
- A nano-network of nanoparticles with a solid core of insulin, modified dextran and glucose oxidase was demonstrated to be able to regulate and maintain normal blood sugar level in a mouse model with Type 1 diabetes for up to 10 days. The nanoparticle is coated with either the positively charged chitosan (from shrimp shells) or negatively charged alginate (from seaweed). The opposite charges hold the nanoparticles together as a nano-network even after they are injected subcutaneously. The porous particles allow blood to make contact with the nanoparticle cores. When there is high blood sugar level, glucose oxidase is activated

and breaks down the dextran into gluconic acid. The disintegration of the cores releases the insulin which controls the blood sugar level to the normal range.

Cancer

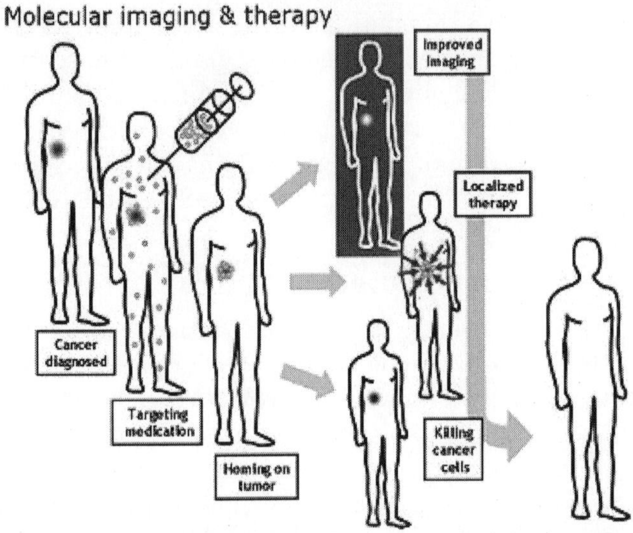

Figure: *A schematic illustration showing how nanoparticles or other cancer drugs might be used to treat cancer.*

The small size of nanoparticles endows them with properties that can be very useful in oncology, particularly in imaging. Quantum dots (nanoparticles with quantum confinement properties, such as size-tunable light emission), when used in conjunction with MRI (magnetic resonance imaging), can produce exceptional images of tumor sites. These nanoparticles are much brighter than organic dyes and only need one light source for excitation. This means that the use of fluorescent quantum dots could produce a higher contrast image and at a lower cost than today's organic dyes used as contrast media. The downside, however, is that quantum dots are usually made of quite toxic elements.

Another nanoproperty, high surface area to volume ratio, allows many functional groups to be attached to a nanoparticle, which can seek out and bind to certain tumor cells. Additionally, the small size of nanoparticles (10 to 100 nanometres), allows them to preferentially accumulate at tumor sites (because tumors lack an effective lymphatic drainage system). Research into multifunctional nanoparticles that would detect, image, and then proceed to treat a tumor is under way. A promising new cancer treatment that may one day replace radiation and chemotherapy is edging closer to human trials. Kanzius RF therapy attaches microscopic nanoparticles to cancer cells and then "cooks" tumors inside the body with radio waves that heat only the nanoparticles and the adjacent (cancerous) cells.

Sensor test chips containing thousands of nanowires, able to detect proteins and other biomarkers left behind by cancer cells, could enable the detection and diagnosis of cancer in the early stages from a few drops of a patient's blood.

The basic point to use drug delivery is based upon three facts: a) efficient encapsulation of the drugs, b) successful delivery of said drugs to the targeted region of the body, and c) successful release of that drug there.

Researchers at Rice University under Prof. Jennifer West, have demonstrated the use of 120 nm diameter nanoshells coated with gold to kill cancer tumors in mice. The nanoshells can be targeted to bond to cancerous cells by conjugating antibodies or peptides to the nanoshell surface. By irradiating the area of the tumor with an infrared laser, which passes through flesh without heating it, the gold is heated sufficiently to cause death to the cancer cells.

Limitations to conventional cancer chemotherapy include drug resistance, lack of selectivity, and lack of solubility. Nanoparticles have the potential to overcome these problems. Nanoparticles of cadmium selenide (quantum dots) glow when exposed to ultraviolet light. When injected, they seep into cancer tumors. The surgeon can see the glowing tumor, and use it as a guide for more accurate tumor removal.

In photodynamic therapy, a particle is placed within the body and is illuminated with light from the outside. The light gets absorbed by the particle and if the particle is metal, energy from the light will heat the particle and surrounding tissue. Light may also be used to produce high energy oxygen molecules which will chemically react with and destroy most organic molecules that are next to them (like tumors). This therapy is appealing for many reasons. It does not leave a "toxic trail" of reactive molecules throughout the body (chemotherapy) because it is directed where only the light is shined and the particles exist. Photodynamic therapy has potential for a noninvasive procedure for dealing with diseases, growth and tumors.

Also, gold nanoparticles have the potential to join numerous therapeutic functions into a single platform, by targeting specific tumor cells, tissues and organs. Actually, Conde et al reported the evaluation of the inflammatory response and therapeutic siRNA silencing via RGD-nanoparticles in a lung cancer mouse model. This study reported the use of siRNA/RGD gold nanoparticles capable of targeting tumor cells in two lung cancer xenograft mouse models, resulting in successful and significant *c-Myc* oncogene downregulation followed by tumor growth inhibition and prolonged survival of the animals. This delivery system can achieve translocation of siRNA duplexes directly into the tumour cell cytoplasm and accomplish successful silencing of an oncogene expression. Actually, RGD/siRNA-AuNPs can target preferentially and be taken up by tumor cells via integrin ávâ3-receptor-mediated endocytosis with no cytotoxicity, showing that can accumulate in tumor tissues overexpressing ává3 integrins and selectively delivered *c-Myc* siRNA to suppress tumor growth and angiogenesis.

Surgery

At Rice University, a flesh welder is used to fuse two pieces of chicken meat into a single piece. The two pieces of chicken are placed together touching. A greenish liquid containing gold-coated nanoshells is dribbled along the seam. An infrared laser is traced along the seam, causing the two sides to weld together. This could solve the difficulties and blood leaks caused when the surgeon tries to restitch the arteries that have been cut during a kidney or heart transplant. The flesh welder could weld the artery perfectly.

Visualization

Tracking movement can help determine how well drugs are being distributed or how substances are metabolized. It is difficult to track a small group of cells throughout the body, so scientists used to dye the cells. These dyes needed to be excited by light of a certain wavelength in order for them to light up. While different colour dyes absorb different frequencies of light, there was a need for as many light sources as cells. A way around this problem is with luminescent tags. These tags are quantum dots attached to proteins that penetrate cell membranes. The dots can be random in size, can be made of bio-inert material, and they demonstrate the nanoscale property that colour is size-dependent. As a result, sizes are selected so that the frequency of light used to make a group of quantum dots fluoresce is an even multiple of the frequency required to make another group incandesce. Then both groups can be lit with a single light source. They have also found a way to insert nanoparticles into the affected parts of the body so that those parts of the body will glow showing the tumor growth or shrinkage or also organ trouble.

Tissue Engineering

Nanotechnology may be able to help reproduce or repair damaged tissue. "Tissue engineering" makes use of artificially stimulated cell proliferation by using suitable nanomaterial-based scaffolds and growth factors. For example, bones could be regrown on carbon nanotube scaffolds. Tissue engineering might replace today's conventional treatments like organ transplants or artificial implants. Advanced forms of tissue engineering may lead to life extension. Artificial bone composites are also being manufactured from nanocrystalline calcium phosphates.

Antimicrobial Effects

Nanoparticles can be used in combination therapy for decreasing antibiotic resistance or for their antimicrobial properties. It has been shown that Zinc

Oxide nanoparticles can decrease the antibiotic resistance and enhance the antibacterial activity of Ciprofloxacin against microorganism in Vitro. Nanoparticles can interfere with the different proteins which are interacting in the antibiotic resistance or pharmacologic mechanisms of drugs. Another example is Silver which has been used for the treatment of burns and chronic wounds. Antibacterial and antifungal properties of nanoparticles have widened their applications in medicine and dentistry.

Immune Response

Buckyballs have been investigated for the ability to "interrupt" the allergy/ immune response by preventing mast cells (which cause allergic response) from releasing histamine into the blood and tissues, by binding to free radicals "dramatically better than any anti-oxidant currently available, such as vitamin E". In February 2013, University of Pennsylvania researchers developed a way to prevent nano particles from triggering an immune system response ; which could aid the delivery of medicinal nanoparticles some time in future. This is done by using the protein CD47, found on most mammalian cell membranes which binds to the SIRPa in macrophages; preventing an immune response from being triggered.

Arthroscope

Nanotechnology is helping to advance the use of arthroscopes, which are pencil-sized devices that are used in surgeries with lights and cameras so surgeons can do the surgeries with smaller incisions. The smaller the incisions the faster the healing time which is better for the patients. It is also helping to find a way to make an arthroscope smaller than a strand of hair.

Diagnostic and Medical Devices

Nanotechnology-on-a-chip is one more dimension of lab-on-a-chip technology. Magnetic nanoparticles, bound to a suitable antibody, are used to label specific molecules, structures or microorganisms. Gold nanoparticles tagged with short segments of DNA can be used for detection of genetic sequence in a sample. Multicolour optical coding for biological assays has been achieved by embedding different-sized quantum dots into polymeric microbeads. Nanopore technology for analysis of nucleic acids converts strings of nucleotides directly into electronic signatures.

- C-dots (Cornell dots) are the smallest silica-based nanoparticles with the size <10 nm. The particles are infused with organic dye which will light up with fluorescence. Clinical trial is underway since 2011 to use the C-dots as diagnostic tool to assist surgeons to identify the location of tumor cells.

Nanotechnology is also opening up new opportunities in implantable delivery systems, which are often preferable to the use of injectable drugs, because the latter frequently display first-order kinetics (the blood concentration goes up rapidly, but drops exponentially over time). This rapid rise may cause difficulties with toxicity, and drug efficacy can diminish as the drug concentration falls below the targeted range.

Neuro-Electronic Interfaces

Neuro-electronic interfacing is a visionary goal dealing with the construction of nanodevices that will permit computers to be joined and linked to the nervous system. This idea requires the building of a molecular structure that will permit control and detection of nerve impulses by an external computer. The computers will be able to interpret, register, and respond to signals the body gives off when it feels sensations. The demand for such structures is huge because many diseases involve the decay of the nervous system (ALS and multiple sclerosis). Also, many injuries and accidents may impair the nervous system resulting in dysfunctional systems and paraplegia. If computers could control the nervous system through neuro-electronic interface, problems that impair the system could be controlled so that effects of diseases and injuries could be overcome. Two considerations must be made when selecting the power source for such applications. They are refuelable and nonrefuelable strategies. A refuelable strategy implies energy is refilled continuously or periodically with external sonic, chemical, tethered, magnetic, or electrical sources. A nonrefuelable strategy implies that all power is drawn from internal energy storage which would stop when all energy is drained.

One limitation to this innovation is the fact that electrical interference is a possibility. Electric fields, Electromagnetic pulses (EMP), and stray fields from other *in vivo* electrical devices can all cause interference. Also, thick insulators are required to prevent electron leakage, and if high conductivity of the *in vivo* medium occurs there is a risk of sudden power loss and "shorting out." Finally, thick wires are also needed to conduct substantial power levels without overheating. Little practical progress has been made even though research is happening. The wiring of the structure is extremely difficult because they must be positioned precisely in the nervous system so that it is able to monitor and respond to nervous signals. The structures that will provide the interface must also be compatible with the body's immune system so that they will remain unaffected in the body for a long time. In addition, the structures must also sense ionic currents and be able to cause currents to flow backward. While the potential for these structures is amazing, there is no timetable for when they will be available.

Medical Applications of Molecular Nanotechnology

Molecular nanotechnology is a speculative subfield of nanotechnology regarding the possibility of engineering molecular assemblers, machines which could re-order matter at a molecular or atomic scale. Molecular nanotechnology is highly theoretical, seeking to anticipate what inventions nanotechnology might yield and to propose an agenda for future inquiry. The proposed elements of molecular nanotechnology, such as molecular assemblers and nanorobots are far beyond current capabilities.

Nanorobots

The somewhat speculative claims about the possibility of using nanorobots in medicine, advocates say, would totally change the world of medicine once it is realised. Nanomedicine would make use of these nanorobots (e.g., Computational Genes), introduced into the body, to repair or detect damages and infections. According to Robert Freitas of the Institute for Molecular Manufacturing, a typical blood borne medical nanorobot would be between 0.5–3 micrometres in size, because that is the maximum size possible due to capillary passage requirement. Carbon could be the primary element used to build these nanorobots due to the inherent strength and other characteristics of some forms of carbon (diamond/fullerene composites), and nanorobots would be fabricated in desktop nanofactories specialised for this purpose.

Nanodevices could be observed at work inside the body using MRI, especially if their components were manufactured using mostly ^{13}C atoms rather than the natural ^{12}C isotope of carbon, since ^{13}C has a nonzero nuclear magnetic moment. Medical nanodevices would first be injected into a human body, and would then go to work in a specific organ or tissue mass. The doctor will monitor the progress, and make certain that the nanodevices have gotten to the correct target treatment region. The doctor will also be able to scan a section of the body, and actually see the nanodevices congregated neatly around their target (a tumor mass, etc.) so that he or she can be sure that the procedure was successful.

Cell Repair Machines

Using drugs and surgery, doctors can only encourage tissues to repair themselves. With molecular machines, there will be more direct repairs. Cell repair will utilise the same tasks that living systems already prove possible. Access to cells is possible because biologists can insert needles into cells without killing them. Thus, molecular machines are capable of entering the cell. Also, all specific biochemical interactions show that molecular systems can recognise other molecules by touch, build or rebuild every molecule in a cell, and can disassemble damaged molecules. Finally, cells that replicate prove that molecular systems can assemble every system found in a cell. Therefore, since nature has demonstrated the basic operations needed to perform molecular-level cell repair, in the future, nanomachine based systems will be built that are able to enter cells, sense differences from healthy ones and make modifications to the structure. The healthcare possibilities of these cell repair machines are impressive. Comparable to the size of viruses or bacteria, their compact parts would allow them to be more complex. The early machines will be specialised. As they open and close cell membranes or travel through tissue and enter cells and viruses, machines will only be able to correct a single molecular disorder like DNA damage or enzyme deficiency. Later, cell repair machines will be programmed with more abilities with the help of advanced AI systems.

Nanocomputers will be needed to guide these machines. These computers will direct machines to examine, take apart, and rebuild damaged molecular structures. Repair machines will be able to repair whole cells by working structure by structure. Then by working cell by cell and tissue by tissue, whole organs can be repaired.

Finally, by working organ by organ, health is restored to the body. Cells damaged to the point of inactivity can be repaired because of the ability of molecular machines to build cells from scratch. Therefore, cell repair machines will free medicine from reliance on self repair alone.

Nanonephrology

Nanonephrology is a branch of nanomedicine and nanotechnology that seeks to use nanomaterials and nanodevices for the diagnosis, therapy, and management of renal diseases. It includes the following goals:

1. the study of kidney protein structures at the atomic level
2. nano-imaging approaches to study cellular processes in kidney cells
3. nanomedical treatments that utilise nanoparticles to treat various kidney diseases

Advances in Nanonephrology are expected to be based on discoveries in the above areas that can provide nano-scale information on the cellular molecular machinery involved in normal kidney processes and in pathological states. By understanding the physical and chemical properties of proteins and other macromolecules at the atomic level in various cells in the kidney, novel therapeutic approaches can be designed to combat major renal diseases. The nanoscale artificial kidney is a goal that many physicians dream of. Nanoscale engineering advances will permit programmable and controllable nano-scale robots to execute curative and reconstructive procedures in the human kidney at the cellular and molecular levels. Designing nanostructures compatible with the kidney cells and that can safely operate in vivo is also a future goal. The ability to direct events in a controlled fashion at the cellular nanolevel has the potential of significantly improving the lives of patients with kidney diseases.

Photodynamic Therapy

Photodynamic therapy (PDT), sometimes called photochemotherapy, is a form of phototherapy using nontoxic light-sensitive compounds that are exposed selectively to light, whereupon they become toxic to targeted malignant and other diseased cells. PDT has proven ability to kill microbial cells, including bacteria, fungi and viruses.

PDT is popularly used in treating acne. It is used clinically to treat a wide range of medical conditions, including wet age-related macular degeneration and malignant cancers, and is recognised as a treatment strategy which is both minimally invasive and minimally toxic.

Most modern PDT applications involve three key components: a photosensitizer, a light source and tissue oxygen. The combination of these three components leads to the chemical destruction of any tissues which have either selectively taken up the photosensitizer or have been locally exposed to light. The wavelength of the light source needs to be appropriate for exciting the photosensitizer to produce reactive oxygen species.

These reactive oxygen species generated through PDT are free radicals (Type I PDT) generated through electron abstraction or transfer from a substrate molecule and highly reactive state of oxygen known as singlet oxygen (Type II PDT). In understanding the mechanism of PDT it is important to distinguish it from other light-based and laser therapies such as laser wound healing and rejuvenation, or intense pulsed light hair removal, which do not require a photosensitizer.

Procedure

In order to achieve the selective destruction of the target area using PDT while leaving normal tissues untouched, either the photosensitizer can be applied locally to the target area, or photosensitive targets can be locally excited with light. For instance, in the treatment of skin conditions, including acne, psoriasis, and also skin cancers, the photosensitizer can be applied topically and locally excited by a light source. In the local treatment of internal tissues and cancers, after photosensitizers have been administered intravenously, light can be delivered to the target area using endoscopes and fiber optic catheters.

Photosensitizers can also target many viral and microbial species, including HIV and MRSA. Using PDT, pathogens present in samples of blood and bone marrow can be decontaminated before the samples are used further for transfusions or transplants. PDT can also eradicate a wide variety of pathogens of the skin and of the oral cavities. Given the seriousness that drug resistant pathogens have now become, there is increasing research into PDT as a new antimicrobial therapy.

Photosensitizers

In air and tissue, molecular oxygen occurs in a triplet state, whereas almost all other molecules are in a singlet state. Reactions between these are forbidden by quantum mechanics, thus oxygen is relatively non-reactive at physiological conditions. A photosensitizer is a chemical compound that can be promoted to an excited state upon absorption light and undergo intersystem crossing with oxygen to produce singlet oxygen. This species rapidly attacks any organic compounds it encounters, thus being highly cytotoxic. It is rapidly eliminated: in cells, the average lifetime is 3 μs.

A wide array of photosensitizers for PDT exist. They can be divided into porphyrins, chlorophylls and dyes. Some examples include aminolevulinic acid (ALA), Silicon Phthalocyanine Pc 4, m-tetrahydroxyphenylchlorin (mTHPC), and mono-L-aspartyl chlorin e6 (NPe6).

Several photosensitizers are commercially available for clinical use, such as Allumera, Photofrin, Visudyne, Levulan, Foscan, Metvix, Hexvix, Cysview, and Laserphyrin, with others in development, e.g. Antrin, Photochlor, Photosens, Photrex, Lumacan, Cevira, Visonac, BF-200 ALA. Amphinex. Also Azadipyrromethenes.

Although these photosensitizers can be used for wildly different treatments, they all aim to achieve certain characteristics:

- High absorption at long wavelengths
  - Tissue is much more transparent at longer wavelengths (~700–850 nm). Absorbing at longer wavelengths would allow the light to penetrate deeper, and allow the treatment of larger tumors.
- High singlet oxygen quantum yield
- Low photobleaching to prevent degradation of the photosensitizer
- Natural fluorescence
  - Many optical dosimetry techniques, such as fluorescence spectroscopy, depend on the drug being naturally fluorescent
- High chemical stability
- Low dark toxicity
  - The photosensitizer should not be harmful to the target tissue until the treatment beam is applied.
- Preferential uptake in target tissue

The major difference between different types of photosensitizers is in the parts of the cell that they target. Unlike in radiation therapy, where damage is done by targeting cell DNA, most photosensitizers target other cell structures. For example, mTHPC has been shown to localize in the nuclear envelope and do its damage there. In contrast, ALA has been found to localize in the mitochondria and Methylene Blue in the lysosomes.

To allow treatment of deeper tumours some researchers are using internal chemiluminescence to activate the photosensitiser.

PUVA therapy is using psoralen as photosensitiser and UVA ultraviolet as light source, but this form of therapy is usually classified as a separate form of therapy from photodynamic therapy.

Targeted PDT

Some photosensitisers naturally accumulate in the endothelial cells of vascular tissue allowing 'vascular targeted' PDT, but there is also research to target the photosensitiser to the tumour (usually by linking it to antibodies or antibody fragments). It is currently only in pre-clinical studies.

Applications

Compared to normal tissues, most types of cancers are especially active in both the uptake and accumulation of photosensitizers agents, which makes cancers especially vulnerable to PDT. Since photosensitizers can also have a high affinity for vascular endothelial cells.

Usage in Acne

PDT is currently in clinical trials to be used as a treatment for severe acne. Initial results show have shown for it to be effective as a treatment only for severe acne, though some question whether it is better than existing acne treatments. The treatment causes severe redness and moderate to severe pain and burning sensation. A phase II trial, while it showed improvement occurred, failed to show improved response compared to the blue/violet light alone.

History

While the applicability and potential of PDT has been known for over a hundred years, the development of modern PDT has been a gradual one, involving scientific progress in the fields of photobiology and cancer biology, as well as the development of modern photonic devices, such as lasers and LEDs. It was John Toth as product manager for Cooper Medical Devices Corp/ Cooper Lasersonics with early clinical argon dye lasers ca 1981 who acknowledged the "photodynamic chemical effect" of the therapy and wrote the first "white paper" branding the therapy as "Photodynamic Therapy" (PDT) to support efforts in setting up 10 clinical sites in Japan where the term "radiation" had negative connotations.

PDT received even greater interest as result of Thomas Dougherty helping expand clinical trials and forming the International Photodynamic Association, in 1986.

PDT in Ancient Medicine

The earliest recorded treatments that exploited a photosensitizer and a light source, in this case sunlight, for medical effect can be found in ancient Egyptian and Indian sources. Annals over 3000 years old report the use of topically applied vegetable and plant substances to produce photoreactions in skin and cause a repigmentation of depigimented skin lesions, as seen with vitiligo and leukoderma.

The photosensitizing agents used in these ancient therapies have been characterised with modern science as belonging to the psoralen family of chemicals. Psoralens are still in use today in PDT regimes to treat a variety of skin conditions, including vitiligo, psoriasis, neurodermatitis, eczema, cutaneous T-cell lymphoma and lichen ruber planus.

20th-Century Development of PDT

The first detailed scientific evidence that agents, photosensitive synthetic dyes, in combination with a light source and oxygen could have potential therapeutic effect was made at the turn of the 20th century in the laboratory of von Tappeiner in Munich, Germany. Historically this was a time when Germany was leading the world in the industrial synthesis of dyes.

While studying the effects of acridine on paramecia cultures, Oscar Raab, a student of von Tappeiner observed a toxic effect. Fortuitously Raab also observed that light was dependent for the killing of paramecia cultures to take place. Subsequent work in the laboratory of von Tappeiner showed that oxygen was essential for the 'photodynamic action' – a term coined by von Tappeiner.

With the discovery of photodynamic effects, von Tappeiner and colleagues went on to perform the first PDT trial in patients with skin carcinoma using the photosensitizer, eosin, Out of 6 patients with a facial basal cell carcinoma, treated with a 1% eosin solution and a long-term exposure either to sunlight or to arc-lamp light, 4 patients showed total tumour resolution and a relapse-free period of 12 months.

It was only much later, when Thomas Dougherty and co-workers at Roswell Park Cancer Institute, Buffalo NY, clinically tested PDT again. In 1978, they published striking results in which they treated 113 cutaneous or subcutaneous malignant tumors and observed a total or partial resolution of 111 tumors.

The active photosensitizer used in the clinical PDT trial by Dougherty was an agent called Haematoporphyrin Derivative (HpD), which was first characterised in 1960 by Lipson. In his research, Lipson wanted to find a diagnostic agent suitable for the detection of tumours in patients. With the discovery of HpD, Lipson went onto pioneer the use of endoscopes and HpD fluorescence to detect tumours.

As its name suggests, HpD is a porphyrin species derived from haematoporphyrin, Porphyrins have long been considered as suitable agents for tumour photodiagnosis and tumour PDT because cancerous cells exhibit a significantly greater uptake and affinity for porphyrins compared to normal quiescent tissues. This important observation, which underlies the success of PDT to treat cancers, had been established by a number of scientific researchers prior to the discoveries made by Lipson.

In 1924, Policard first revealed the diagnostic capabilities of hematoporphyrin fluorescence when he observed that ultraviolet radiation excited red fluorescence in the sarcomas of laboratory rats. Policard hypothesized at the time that the fluorescence was associated with endogenous hematoporphyrin accumulation.

In 1948, Figge with co-workers showed on laboratory animals that porphyrins exhibit a preferential affinity to rapidly dividing cells, including malignant, embryonic, and regenerative cells, and because of this, they proposed that porphyrins should be used in the treatment of cancer. Subsequently many scientific authors have repeated the observation that cancerous cells naturally accumulate porphyrins and have characterised a number of mechanisms to explain it.

HpD, under the pharmaceutical name Photofrin, was the first PDT agent approved for clinical use in 1993 to treat a form of bladder cancer in Canada. Over the next decade, both PDT and the use of HpD received wider international attention and grew in their clinical use, and lead to the first PDT treatments to receive U.S. Food and Drug Administration approval.

Modern Development of PDT in Russia

Of all the nations beginning to use PDT in the late 20th century, the Russians were the quickest to advance its use clinically and to make many developments. One early Russian development was a new photosensitizer called Photogem which, like HpD, was derived from haematoporphyrin in 1990 by Professor Andrey F. Mironov and coworkers in Moscow. Photogem was approved by the Ministry of Health of Russia and tested clinically from February 1992 to 1996. A pronounced therapeutic effect was observed in 91 percent of the 1500 patients that underwent PDT using Photogem, with 62 percent having a total tumor resolution.

Of the remaining patients, a further 29 percent had a partial tumor resolution, where the tumour at least halved in size. In those patients that had been diagnosed early, 92 percent of the patients showed complete resolution of the tumour.

Around this time, Russian scientists also collaborated with NASA medical scientists who were looking at the use of LEDs as more suitable light sources, compared to lasers, for PDT applications.

Modern Development of PDT in Asia

PDT has also seen considerably development in Asia. Since 1990, the Chinese have been developing specialist clinical expertise with PDT using their own domestically produced photosensitizers, derived from Haematoporphyrin, and light sources. PDT in China is especially notable for the technical skill of specialists in effecting resolution of difficult to reach tumours.

Levulan

Levulan photodynamic therapy is a treatment method for certain skin conditions developed by DUSA Pharmaceuticals. The active ingredient is ä-aminolevulinic acid (ALA).

Actinic Keratosis

U.S. Food and Drug Administration (FDA) approval for actinic keratosis was received in December 1999.

Treatment Protocol

A 20% solution of ALA is applied to the lesions, and after a 6 hour wait to allow proper penetration of the skin, the treatment area is irradiated with blue light for 15 minutes. This generates reactive oxygen species in the skin that can lead to cell death. The treated areas are sensitive to light, and patients must protect themselves against bright sunshine for up to 40 hours after treatment. Patients must stop using topical acne medicine and avoid steam rooms, hot showers, and generally protect the skin for up to one week after treatment.

Acne

Levulan is also prescribed for the treatment of acne but this is an off-label use, which means it is not approved by the FDA.

On October 2008 Dusa Pharmaceuticals announced that it was no longer developing Levulan for the treatment of acne after the phase II results failed to show statistically significant results.

Barrett's Esophagus

A phase II trial was initiated in 2004. In May 2007 the U.S. FDA granted Orphan Drug Designation for Levulan PDT for the treatment of esophageal dysplasia, a complication of Barrett's esophagus.

Impalefection

Impalefection is a method of gene delivery using nanomaterials, such as carbon nanofibers, carbon nanotubes, nanowires Ref.1. Needle-like nanostructures are synthesized perpendicular to the surface of a substrate.

Plasmid DNA containing the gene, intended for intracellular delivery, is attached to the nanostructure surface. A chip with arrays of these needles is then pressed against cells or tissue. Cells that are impaled by nanostructures can express the delivered gene(s). As one of the types of transfection, the term is derived from two words - impalement and infection.

One of the features of impalefection is spatially resolved gene delivery that holds potential for such tissue engineering approaches in wound healing as gene activated matrix technology Ref2.

Vertically aligned carbon nanofiber arrays prepared by photolithography and plasma enhanced chemical vapor deposition are one of the suitable types of material. Silicon nanowires is another choice of nanoneedles that have been utilised for impalefection.

Transfection

Transfection is the process of deliberately introducing nucleic acids into cells. The term is often used for non-viral methods in eukaryotic cells. It may also refer to other methods and cell types, although other terms are preferred: "transformation" is more often used to describe non-viral DNA transfer in bacteria, non-animal eukaryotic cells, including plant cells. In animal cells, transfection is the preferred term as transformation is also used to refer to progression to a cancerous state (carcinogenesis) in these cells. Transduction is often used to describe virus-mediated DNA transfer.

The word *transfection* is a blend of *trans-* and *infection*. Genetic material (such as supercoiled plasmid DNA or siRNA constructs), or even proteins such as antibodies, may be transfected.

Transfection of animal cells typically involves opening transient pores or "holes" in the cell membrane, to allow the uptake of material. Transfection can be carried out using calcium phosphate, by electroporation, by cell squeezing or by mixing a cationic lipid with the material to produce liposomes, which fuse with the cell membrane and deposit their cargo inside.

Transfection can result in unexpected morphologies and abnormalities in target cells.

The meaning of the term has evolved. The original meaning of transfection was "infection by transformation," i.e., introduction of DNA (or RNA) from a prokaryote-infecting virus or bacteriophage into cells, resulting in an infection. Because the term transformation had another sense in animal cell biology (a genetic change allowing long-term propagation in culture, or acquisition of properties typical of cancer cells), the term transfection acquired, for animal cells, its present meaning of a change in cell properties caused by introduction of DNA.

Methods

There are various methods of introducing foreign DNA into a eukaryotic cell: some rely on physical treatment (electroporation, cell squeezing, nanoparticles, magnetofection), other on chemical materials or biological particles (viruses) that are used as carriers.

Chemical-Based Transfection

Chemical-based transfection can be divided into several kinds: cyclodextrin, polymers, liposomes, or nanoparticles (with or without chemical or viral functionalization.).

- One of the cheapest methods uses calcium phosphate, originally discovered by F. L. Graham and A. J. van der Eb in 1973. HEPES-buffered saline solution (HeBS) containing phosphate ions is combined with a calcium chloride solution containing the DNA to be transfected. When the two are combined, a fine precipitate of the positively charged calcium and the negatively charged phosphate will form, binding the DNA to be transfected on its surface. The suspension of the precipitate is then added to the cells to be transfected (usually a cell culture grown in a monolayer). By a process not entirely understood, the cells take up some of the precipitate, and with it, the DNA. This process has been a preferred method of identifying many oncogenes.
- Other methods use highly branched organic compounds, so-called dendrimers, to bind the DNA and get it into the cell.
- A very efficient method is the inclusion of the DNA to be transfected in liposomes, i.e. small, membrane-bounded bodies that are in some ways similar to the structure of a cell and can actually fuse with the cell membrane, releasing the DNA into the cell. For eukaryotic cells, transfection is better achieved using cationic liposomes (or mixtures), because the cells are more sensitive.
- Another method is the use of cationic polymers such as DEAE-dextran or polyethylenimine. The negatively charged DNA binds to the polycation and the complex is taken up by the cell via endocytosis.

Non Chemical Methods

- Electroporation (Gene electrotransfer) is a popular method, where transient increase in the permeability of cell membrane is achieved when the cells are exposed to short pulses of an intense electric field.
- *Cell squeezing* is a novel method invented in 2013 by Dr. Armon Sharei, Prof. Bob Langer and Prof. Klavs Jensen at MIT. It enables delivery of molecules into cells by a gentle squeezing of the cell membrane. It is a high throughput vector-free microfluidic platform for intracellular

delivery. It eliminates the possibility of toxicity or off-target effects as it does not rely on exogenous materials or electrical fields.

- Sonoporation uses high-intensity ultrasound to induce pore formation in cell membranes. This pore formation is attributed mainly to the cavitation of gas bubbles interacting with nearby cell membranes since is enhanced by the addition of ultrasound contrast agent, a source of cavitation nuclei.
- Optical transfection is a method where a tiny (~1 μm diameter) hole is transiently generated in the plasma membrane of a cell using a highly focused laser. This technique was first described in 1984 by Tsukakoshi et al., who used a frequency tripled Nd:YAG to generate stable and transient transfection of normal rat kidney cells. In this technique, one cell at a time is treated, making it particularly useful for single cell analysis.
- Protoplast fusion is a technique in which transformed bacterial cells are treated with lysozyme in order to remove the cell wall. Following this, fusogenic agents (e.g., Sendai virus, PEG, or electroporation) are used in order to fuse the protoplast carrying the gene of interest with the target recipient cell. A major disadvantage of this method is that bacterial components are non-specifically introduced into the target cell as well.
- Impalefection is a method of introducing DNA bound to a surface of a nanofiber that is inserted into a cell. This approach can also be implemented with arrays of nanofibers that are introduced into large numbers of cells and intact tissue.
- Hydrodynamic delivery In mice and rats, but to a lesser extent in larger animals, DNA most often in plasmids, including transposons, can be delivered to the liver using hydrodynamic injection that involves infusion of a relatively large volume in the blood in less than 10 seconds; nearly all of the DNA is expressed in the liver by this procedure.

Particle-Based Methods

- A direct approach to transfection is the gene gun, where the DNA is coupled to a nanoparticle of an inert solid (commonly gold) which is then "shot" directly into the target cell's nucleus.
- Magnetofection, or Magnet assisted transfection is a transfection method, which uses magnetic force to deliver DNA into target cells. Nucleic acids are first associated with magnetic nanoparticles. Then, application of magnetic force drives the nucleic acid particle complexes towards and into the target cells, where the cargo is released.

- Impalefection is carried out by impaling cells by elongated nanostructures and arrays of such nanostructures such as carbon nanofibers or silicon nanowires which have been functionalized with plasmid DNA.
- Another particle-based method of transfection is known as particle bombardment. The nucleic acid is delivered through membrane penetration at a high velocity, usually connected to microprojectiles.

Viral Methods

DNA can also be introduced into cells using viruses as a carrier. In such cases, the technique is called viral transduction, and the cells are said to be transduced. Adenoviral vectors can be useful for viral transfection methods because they can transfer genes into a wide variety of human cells and have high transfer rates.

Other (and Hybrid) Methods

Other methods of transfection include nucleofection, which has proved very efficient in transfection of the THP-1 cell line, creating a viable cell line that was able to be differentiated into mature macrophages, heat shock.

Stable and Transient Transfection

For some applications of transfection, it is sufficient if the transfected genetic material is only transiently expressed. Since the DNA introduced in the transfection process is usually not integrated into the nuclear genome, the foreign DNA will be diluted through mitosis or degraded. Cell lines expressing the Epstein–Barr virus (EBV) nuclear antigen 1 (EBNA1) or the SV40 large-T antigen, allow episomal amplification of plasmids containing the viral EBV (293E) or SV40 (293T) origins of replication, greatly reducing the rate of dilution.

If it is desired that the transfected gene actually remain in the genome of the cell and its daughter cells, a stable transfection must occur. To accomplish this, a marker gene is co-transfected, which gives the cell some selectable advantage, such as resistance towards a certain toxin. Some (very few) of the transfected cells will, by chance, have integrated the foreign genetic material into their genome.

If the toxin is then added to the cell culture, only those few cells with the marker gene integrated into their genomes will be able to proliferate, while other cells will die. After applying this selective stress (selection pressure) for some time, only the cells with a stable transfection remain and can be cultivated further.

A common agent for selecting stable transfection is Geneticin, also known as G418, which is a toxin that can be neutralized by the product of the neomycin resistance gene.

RNA Transfection

RNA can also be transfected into cells to transiently express its coded protein, or to study RNA decay kinetics. The latter application is referred as siRNA transfection or RNA silencing, and has become a major application in research (to replace the "knock-down" experiments, to study the expression of proteins, i.e. of Endothelin-1) with potential applications in gene-therapy.

A limitation of the silencing approach rely on the toxicity of the transfection for cells, and its suspected effect on the expression of other genes/proteins.

Protofection

Protofection is a term that refers to the transfection of foreign mitochondrial DNA into the mitochondria of all cells in a tissue to supplement or replace the native mitochondrial DNA already present. As mitochondrial DNA becomes progressively more damaged with age, this may provide a method of at least partially rejuvenating mitochondria in old tissue, restoring them to their original, youthful function. It is thought that mitochondrial damage and dysfunction play an important role in aging.

Protofection is also a basis for constructing gene therapies for inherited genetic diseases such as Leber's hereditary optic neuropathy in which mitochondrial DNA is mutated.

This technology could similarly be applied to modified or artificial mitochondria. The intent being designing ones that do not produce as many (preferably zero) free radicals while staying as, or more efficient in generating energy in the cell. While not invulnerable to free radical damage, having less free radicals would also make such generators have longer lifespans if they could still renew at an identical rate, or at least enough to keep more healthy ones at a given time.

2

Medical Monitoring

In medicine, monitoring is the observation of a disease, condition or one or several medical parameters over time.

It can be performed by continuously measuring certain parameters by using a medical monitor (for example, by continuously measuring vital signs by a bedside monitor), and/or by repeatedly performing medical tests (such as blood glucose monitoring with a glucose metre in people with diabetes mellitus).

Transmitting data from a monitor to a distant monitoring station is known as telemetry or biotelemetry.

Classification by Target Parameter

Monitoring can be classified by the target of interest, including:

- Cardiac monitoring, which generally refers to continuous electrocardiography with assessment of the patients condition relative to their cardiac rhythm.

  A small monitor worn by an ambulatory patient for this purpose is known as a Holter monitor. Cardiac monitoring can also involve cardiac output monitoring via an invasive Swan-Ganz catheter.
- Hemodynamic monitoring, which monitors the blood pressure and blood flow within the circulatory system. Blood pressure can be measured either invasively through an inserted blood pressure transducer assembly, or noninvasively with an inflatable blood pressure cuff.

- Respiratory monitoring, such as:
    - o Pulse oximetry which involves measurement of the saturated percentage of oxygen in the blood, referred to as SpO2, and measured by an infrared finger cuff
    - o Capnography, which involves CO_2 measurements, referred to as EtCO2 or end-tidal carbon dioxide concentration. The respiratory rate monitored as such is called AWRR or airway respiratory rate)
    - o Respiratory rate monitoring through a thoracic transducer belt, an ECG channel or via capnography
- Neurological monitoring, such as of intracranial pressure. Also, there are special patient monitors which incorporate the monitoring of brain waves (electroencephalography), gas anesthetic concentrations, bispectral index (BIS), etc. They are usually incorporated into anesthesia machines. In neurosurgery intensive care units, brain EEG monitors have a larger multichannel capability and can monitor other physiological events, as well.
- Blood glucose monitoring
- Childbirth monitoring
- Body temperature monitoring through an adhesive pad containing a thermoelectric transducer.

Vital Parameters

Monitoring of vital parameters can include several of the ones mentioned above, and most commonly include at least blood pressure and heart rate, and preferably also pulse oximetry and respiratory rate. Multimodal monitors that simultaneously measure and display the relevant vital parameters are commonly integrated into the bedside monitors in critical care units, and the anesthetic machines in operating rooms.

These allow for continuous monitoring of a patient, with medical staff being continuously informed of the changes in general condition of a patient. Some monitors can even warn of pending fatal cardiac conditions before visible signs are noticeable to clinical staff, such as atrial fibrillation or premature ventricular contraction (PVC).

Medical Monitor

A *medical monitor* or *physiological monitor* is a medical device used for monitoring. It can consist of one or more sensors, processing components, display devices (which are sometimes in themselves called "monitors"), as well as communication links for displaying or recording the results elsewhere through a monitoring network.

Components

Sensor: Sensors of medical monitors include biosensors and mechanical sensors.

Translating Component

The translating component of medical monitors is responsible for converting the signals from the sensors to a format that can be shown on the display device or transferred to an external display or recording device.

Display Device

Physiological data are displayed continuously on a CRT, LED or LCD screen as data channels along the time axis, They may be accompanied by numerical readouts of computed parameters on the original data, such as maximum, minimum and average values, pulse and respiratory frequencies, and so on. Besides the tracings of physiological parameters along time (X axis), digital medical displays have automated numeric readouts of the peak and/ or average parameters displayed on the screen.

Modern medical display devices commonly use digital signal processing (DSP), which has the advantages of miniaturization, portability, and multi-parameter displays that can track many different vital signs at once.

Old analog patient displays, in contrast, were based on oscilloscopes, and had one channel only, usually reserved for electrocardiographic monitoring (ECG). Therefore, medical monitors tended to be highly specialised. One monitor would track a patient's blood pressure, while another would measure pulse oximetry, another the ECG. Later analog models had a second or third channel displayed in the same screen, usually to monitor respiration movements and blood pressure. These machines were widely used and saved many lives, but they had several restrictions, including sensitivity to electrical interference, base level fluctuations and absence of numeric readouts and alarms.

Communication Links

Several models of multi-parameter monitors are networkable, i.e., they can send their output to a central ICU monitoring station, where a single staff member can observe and respond to several bedside monitors simultaneously. Ambulatory telemetry can also be achieved by portable, battery-operated models which are carried by the patient and which transmit their data via a wireless data connection.

Digital monitoring has created the possibility, which is being fully developed, of integrating the physiological data from the patient monitoring networks into the emerging hospital electronic health record and digital charting systems, using appropriate health care standards which have been developed for this purpose by organisations such as IEEE and HL7. This newer method of charting patient data reduces the likelihood of human documentation error and will eventually reduce overall paper consumption. In addition, automated ECG interpretation incorporates diagnostic codes automatically into the charts. Medical monitor's embedded software can take care of the data coding according to these standards and send messages to the medical records application, which decodes them and incorporates the data into the adequate fields.

Long-distance connectivity can avail for telemedicine, which involves provision of clinical health care at a distance.

Other Components

A medical monitor can also have the function to produce an alarm (such as using audible signals) to alert the staff when certain criteria are set, such as when some parameter exceeds of falls the level limits.

Mobile Appliances

An entirely new scope is opened with mobile carried monitors, even such in sub-skin carriage. This class of monitors delivers information gathered in body-area networking (BAN) to e.g. smart phones and implemented autonomous agents.

Interpretation of Monitored Parameters

Monitoring of clinical parameters is primarily intended to detect changes (or absence of changes) in the clinical status of an individual. For example, the parameter of oxygen saturation is usually monitored to detect changes in respiratory capability of an individual.

Change in Status versus Test Variability

When monitoring a clinical parameters, differences between test results (or values of a continuously monitored parameter after a time interval) can reflect either (or both) an actual change in the status of the condition or a test-retest variability of the test method.

In practice, the possibility that a difference is due to test-retest variability can almost certainly be excluded if the difference is larger than a predefined "critical difference". This "critical difference" (CD) is calculated as:

$$CD = K \times \sqrt{CV_a^2 + CV_i^2}$$

where:

- K, is a factor dependent on the preferred probability level. Usually, it is set at 2.77, which reflects a 95% prediction interval, in which case there is less than 5% probability that a test result would become higher or lower than the critical difference by test-retest variability in the absence of other factors.

- CV_a is the analytical variation
- CV_i is the intra-individual variability

For example, if a patient has a hemoglobin level of 100 g/L, the anaytical variation (CV_a) is 1.8% and the intra-individual variability CV_i is 2.2%, then the critical difference is 8.1 g/L.

Thus, for changes of less than 8 g/L since a previous test, the possibility that the change is completely caused by test-retest variability may need to be considered in addition to considering effects of, for example, diseases or treatments.

Critical Differences for Some Blood Tests	
Sodium	3%
Potassium	14%
Chloride	4%
Urea	30%
Creatinine	14%
Calcium	5%
Albumin	8%
Fasting glucose	15%
Amylase	30%
Carcinoembryonic antigen	69%
C-reactive protein	43%
Glycosylated hemoglobin	21%
Hemoglobin	8%
Erythrocytes	10%
Leukocytes	32%
Platelets	25%

Unless otherwise specified, then reference for critical values is *Fraser 1989*

Critical differences for other tests include early morning urinary albumin concentration, with a critical difference of 40%.

Techniques in Development

The development of new techniques for monitoring is an advanced and developing field in smart medicine, biomedical-aided integrative medicine, alternative medicine, self-tailored preventive medicine and predictive medicine that emphasizes monitoring of comprehensive medical data of patients, people at risk and healthy people using advanced, smart, minimally invasive biomedical devices, biosensors, lab-on-a-chip (in the future nanomedicine devices like nanorobots) and advanced computerized medical diagnosis and early warning tools over a short clinical interview and drug prescription.

As biomedical research, nanotechnology and nutrigenomics advances, realising the human body's self-healing capabilities and the growing awareness of the limitations of medical intervention by chemical drugs-only approach of old school medical treatment, new researches that shows the enormous damage medications can cause, researchers are working to fulfill the need for a comprehensive further study and personal continuous clinical monitoring of health conditions while keeping legacy medical intervention as a last resort.

In many medical problems, drugs offer temporary relief of symptoms while the root of a medical problem remains unknown without enough data of all our biological systems . Our body is equipped with sub-systems for the purpose of maintaining balance and self healing functions. Intervention without sufficient data might damage those healing sub systems. Monitoring medicine fills the gap to prevent diagnosis errors and can assist in future medical research by analyzing all data of many patients.

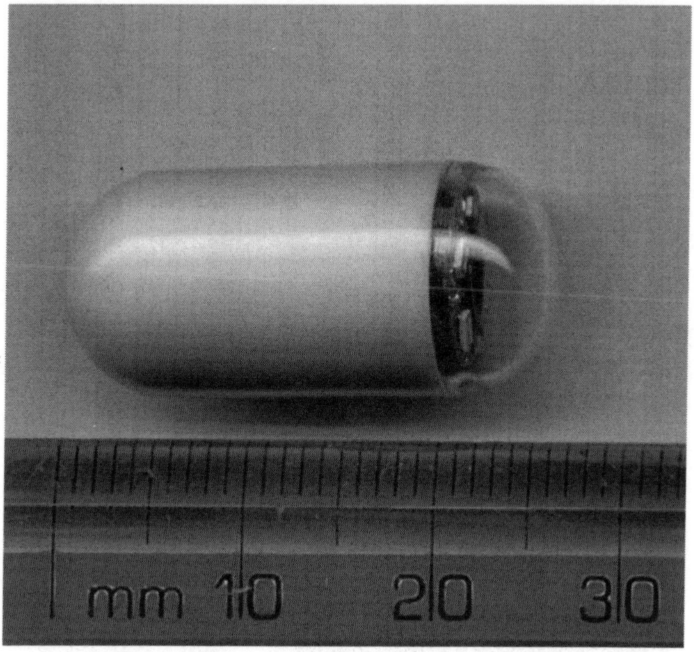

Figure: *Given Imaging Capsule endoscopy*

Examples and Applications

The development cycle in medicine is extremely long, up to 20 years, because of the need for U.S. Food and Drug Administration (FDA) approvals, therefore many of monitoring medicine solutions are not available today in conventional medicine.

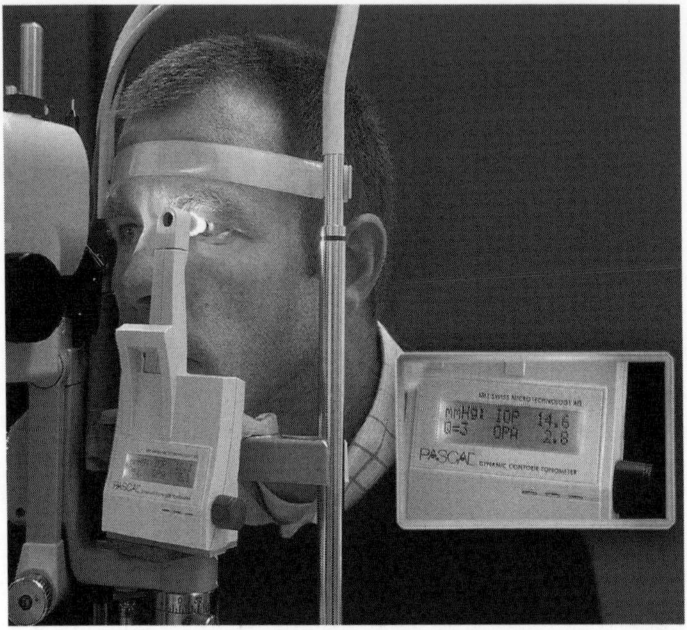

Figure: *The PASCAL Dynamic Contour Tonometre. A monitor for detection of increased intraocular pressure.*

Blood Glucose Monitoring: In vivo blood glucose monitoring devices can transmit data to a computer that can assist with daily life suggestions for lifestyle or nutrition and with the physician can make suggestions for further study in people who are at risk and help prevent diabetes mellitus type 2 .

Stress Monitoring: Bio sensors may provide warnings when stress levels signs are rising before human can notice it and provide alerts and suggestions.

Serotonin Biosensor: Future serotonin biosensors may assist with mood disorders and depression.

Continuous Blood Test Based Nutrition: In the field of evidence-based nutrition, a lab-on-a-chip implant that can run 24/7 blood tests may provide a continuous results and a coumputer can provide nutritaion suggestions or alerts.

Psychiatrist-on-a-Chip: In clinical brain sciences drug delivery and in vivo Bio-MEMS based biosensors may assist with preventing and early treatment of mental disorders

Epilepsy Monitoring: In epilepsy, next generations of long-term video-EEG monitoring may predict epileptic seizure and prevent them with changes of daily life activity like sleep, stress, nutrition and mood management.

Toxicity Monitoring: Smart biosensors may detect toxic materials such mercury and lead and provide alerts.

Particle Therapy

Particle therapy is a form of external beam radiotherapy using beams of energetic protons, neutrons, or positive ions for cancer treatment. The most common type of particle therapy as of 2012 is proton therapy. Although a photon, used in x-ray or gamma ray therapy, can also be considered a particle, photon therapy is not considered here. Additionally, electron therapy is generally put into its own category. Because of this, particle therapy is sometimes referred to, more correctly, as hadron therapy (that is, therapy with particles that are made of quarks).

Neutron capture therapy might be considered a type of particle therapy, but it is not discussed here, as the damage it does to tumors is mostly from energetic ions produced by the secondary nuclear reaction after the neutrons in the external beam are absorbed into boron-10 (or occasionally some other nuclide), and not due primarily to the neutrons themselves. It is therefore a type of secondary particle therapy.

Muon therapy, a rare type of particle therapy not within the categories above, has occasionally been attempted.

Method

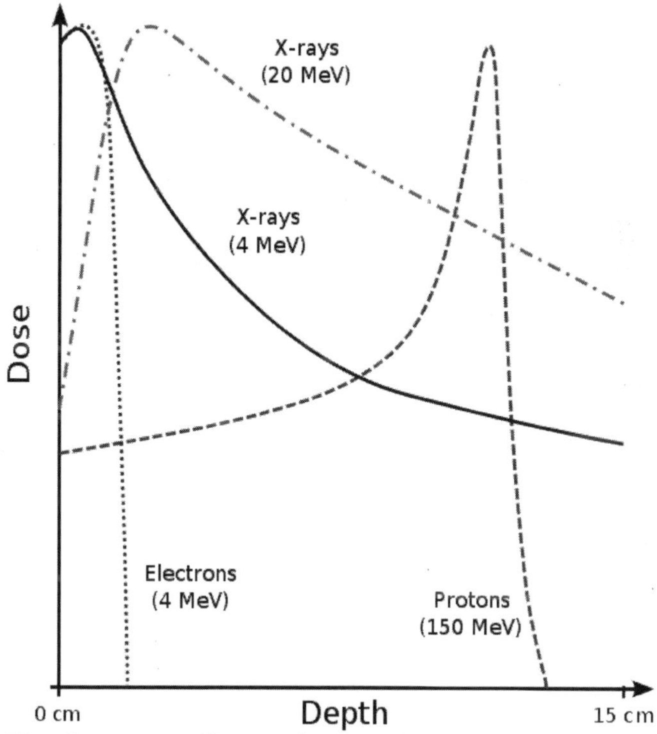

Figure: *Unlike electrons or X-rays, the dose from protons to tissue is maximum just over the last few millimetres of the particle's range.*

Particle therapy works by aiming energetic ionizing particles at the target tumor. These particles damage the DNA of tissue cells, ultimately causing their death. Because of their reduced ability to repair damaged DNA, cancerous cells are particularly vulnerable to attack. The figure shows how beams of electrons, X-rays or protons of different energies (expressed in MeV) penetrate human tissue. Electrons have a short range and are therefore only of interest close to the skin. Bremsstrahlung X-rays penetrate more deeply, but the dose absorbed by the tissue then shows the typical exponential decay with increasing thickness. For protons and heavier ions, on the other hand, the dose increases while the particle penetrates the tissue and loses energy continuously. Hence the dose increases with increasing thickness up to the Bragg peak that occurs near the end of the particle's range. Beyond the Bragg peak, the dose drops to zero (for protons) or almost zero (for heavier ions).

The advantage of this energy deposition profile is that less energy is deposited into the healthy tissue surrounding the target tissue.

The ions are first accelerated by means of a cyclotron or synchrotron. The final energy of the emerging particle beam defines the depth of penetration, and hence, the location of the maximum energy deposition. Since it is easy to deflect the beam by means of electro-magnets in a transverse direction, it is possible to employ a raster scan method, i.e., to scan the target area quickly like the electron beam scans a TV tube. If, in addition, the beam energy and hence, the depth of penetration is varied, an entire target volume can be covered in three dimensions, providing an irradiation exactly following the shape of the tumor. This is one of the great advantages compared to conventional X-ray therapy.

At the end of 2008, 28 treatment facilities were in operation worldwide and over 70,000 patients had been treated by means of pions, protons and heavier ions. Most of this therapy has been conducted using protons.

Heavy-Ion Therapy

Heavy-ion therapy is the use of particles more massive than protons or neutrons, such as carbon ions. Compared to protons, carbon ions have an advantage: due to the higher density of ionisation at the end of their range, correlated damages of the DNA structure within one cell occur more often so that it becomes more difficult for the cancerous cell to repair the damage. This increases the biological efficiency of the dose by a factor between 1.5 and 3. Compared to protons, carbon ions have the disadvantage that beyond the Bragg peak, the dose does not decrease to zero, since nuclear reactions between the carbon ions and the atoms of the tissue lead to production of lighter ions which have a higher range. Therefore, some damage occurs also beyond the Bragg peak.

By the end of 2008, more than 5,000 patients had been treated using carbon ions.

Carbon-Ion Treatment Centres in Operation

The Particle Therapy Co-Operative Group continuously lists treatment centres in operation and in the planning or construction stage. Three centres using carbon ions are in operation, two in Japan: the HIMAC at Chiba and the HIBMC at Hyogo. In Germany, treatment at the *Gesellschaft für Schwerionenforschung* (GSI) in Darmstadt, which is primarily a physics laboratory, has been discontinued in 2008, but the new HIT in Heidelberg, which is a dedicated facility, started in November 2009. The CNAO in Pavia, Italy is going to open in 2011 and will be one of the most advanced centres for particle therapy with hadrons. CNAO will combine precise dose delivery with highly accurate patient alignment based on stereoscopic X-ray imaging. Sophisticated approaches in image-guided particle therapy (IGPT) augments the radiotherapy machines with imaging capabilities and the latest computer vision technology to increase accuracy of target localization and enable patient alignment accuracies of 0.5 mm and better.

Top-Down and Bottom-Up Design

Top-down and bottom-up are both strategies of information processing and knowledge ordering, used in a variety of fields including software, humanistic and scientific theories, and management and organisation. In practice, they can be seen as a style of thinking and teaching.

A top-down approach (also known as stepwise design and in some cases used as a synonym of *decomposition*) is essentially the breaking down of a system to gain insight into its compositional sub-systems. In a top-down approach an overview of the system is formulated, specifying but not detailing any first-level subsystems. Each subsystem is then refined in yet greater detail, sometimes in many additional subsystem levels, until the entire specification is reduced to base elements. A top-down model is often specified with the assistance of "black boxes", these make it easier to manipulate. However, black boxes may fail to elucidate elementary mechanisms or be detailed enough to realistically validate the model. Top down approach starts with the big picture. It breaks down from there into smaller segments.

A bottom-up approach is the piecing together of systems to give rise to more complex systems, thus making the original systems sub-systems of the emergent system. Bottom-up processing is a type of information processing based on incoming data from the environment to form a perception. Information enters the eyes in one direction (input), and is then turned into an image by the brain that can be interpreted and recognised as a perception (output). In a bottom-up approach the individual base elements of the system are first specified in great detail. These elements are then linked together to form larger subsystems, which then in turn are linked, sometimes in many levels, until a complete top-level system is formed. This strategy often resembles a "seed" model, whereby the beginnings are small but eventually grow in complexity and completeness.

However, "organic strategies" may result in a tangle of elements and subsystems, developed in isolation and subject to local optimization as opposed to meeting a global purpose.

Top-down and bottom-up are two approaches for the manufacture of products. These terms were first applied to the field of nanotechnology by the Foresight Institute in 1989 in order to distinguish between molecular manufacturing (to mass-produce large atomically precise objects) and conventional manufacturing (which can mass-produce large objects that are not atomically precise). Bottom-up approaches seek to have smaller (usually molecular) components built up into more complex assemblies, while top-down approaches seek to create nanoscale devices by using larger, externally controlled ones to direct their assembly.

The top-down approach often uses the traditional workshop or microfabrication methods where externally controlled tools are used to cut, mill, and shape materials into the desired shape and order. Micropatterning techniques, such as photolithography and inkjet printing belong to this category.

Bottom-up approaches, in contrast, use the chemical properties of single molecules to cause single-molecule components to (a) self-organise or self-assemble into some useful conformation, or (b) rely on positional assembly. These approaches utilise the concepts of molecular self-assembly and/or molecular recognition. Such bottom-up approaches should, broadly speaking, be able to produce devices in parallel and much cheaper than top-down methods, but could potentially be overwhelmed as the size and complexity of the desired assembly increases.

Medical Uses of Silver

The medical uses of silver include its incorporation into wound dressings, creams, and as an antibiotic coating on medical devices. While wound dressings containing silver sulfadiazine or silver nanomaterials may be used on external infections, there is little evidence to support such use. There is tentative evidence that silver coatings on urinary catheters and endotracheal breathing tubes may reduce the incidence of catheter-related urinary tract infections and ventilator-associated pneumonia, respectively.

The silver ion (Ag+) is bioactive and in sufficient concentration readily kills bacteria *in vitro*. Silver exhibits low toxicity in the human body, and minimal risk is expected due to clinical exposure by inhalation, ingestion, dermal application. Silver and silver nanoparticles are used as an antimicrobial in a variety of industrial, healthcare and domestic applications.

Colloidal silver (a colloid consisting of silver particles suspended in liquid) and formulations containing silver salts were used by physicians in the early 20th century, but their use was largely discontinued in the 1940s following the development of safer and effective modern antibiotics. Since the 1990s, colloidal silver has again been marketed as an alternative medicine, often with extensive "cure-all" claims. Colloidal silver products remain available in many countries as dietary supplements and homeopathic remedies, although they are not effective in treating any known condition and carry the risk of both permanent cosmetic side effects such as argyria and more serious ones such as allergic reactions, and interactions with prescription medications.

Medical Uses

Antibacterial Cream: A 2012 systematic review reported that topical silver showed significantly worse healing time compared to controls and showed no evidence of effectiveness in preventing wounds infection. A 2010 Cochrane systematic review concluded that "There is insufficient evidence to establish whether silver-containing dressings or topical agents promote wound healing or prevent wound infection".

The US Food and Drug Administration has approved a number of topical preparations of silver sulfadiazine for treatment of second- and third-degree burns.

Silver Sulfadiazine: Silver sulfadiazine (INN, or silvadene) is a topical sulfonamide/silver antibacterial used as a topical cream on burns. Studies have found that silver sulfadiazine may increase healing times – wounds may take longer to heal if treated with this drug and it is therefore not recommended by the authors of a Cochrane review.

Silver sulfadiazine is typically delivered in a 1% cream or aqueous suspension. Brand names include Silvadene (a genericized trademark), Silverex, Silverol, Silvazine, Flamazine, Thermazene, BurnHeal, and SSD.

It is on the World Health Organisation's List of Essential Medicines, a list of the most important medication needed in a basic health system.

A 2012 review found that topical silver worsens healing time compared to controls and showed no evidence of effectiveness in preventing wounds infection. A Cochrane systematic review from 2008 found that most of the trials had methodological shortcomings and thus are of little use. Another Cochrane systematic review from 2010 concluded that "There is insufficient evidence to establish whether silver-containing dressings or topical agents promote wound healing or prevent wound infection". Other reviews of the evidence have also concluded that the "quality of the trials was limited". Cochrane has raised concerns about delays in time to wound healing and an increased number of dressing applications when SSD is used.

Adverse Effects: An aseptic exudate (a clear fluid) may form on the wound's surface. Burning and painful sensations are not uncommon but only temporary. About 0.1 to 1% of patients show hypersensitivity reactions like rashes or erythema multiforme. This is a reaction known from other sulfonamides including antibacterials, thiazide diuretics, and sulfonylurea antidiabetics; but data on the likelihood of cross-allergies are inconsistent.

Incorporation of the silver ions can lead to local argyria (discolouration of the skin), especially if the treated area is exposed to ultraviolet. Generalised argyria with silver accumulation in kidneys, liver and retina has only been found in association with excessive long-term use, or repeated use on severe and heavily inflamed burns. Possible consequences of generalised argyria include interstitial nephritis and anemia.

Interactions

Proteases like trypsin and clostridiopeptidase, which are contained in ointments used for the removal of dead skin on wounds, can be inhibited by silver ions if applied simultaneously. When silver sulfadiazine is absorbed in significant amounts, it can increase effects and side effects of some drugs like vitamin K antagonist.

Pharmacokinetics

The chemical itself is poorly soluble, and has only very limited penetration through the skin. Only when applied to large area burns or other lesions is absorption into the body a problem.

Dressings

A 2012 systematic review found that silver-containing dressings were no better than non-silver-containing dressings in treating burns. A 2012 Cochrane review found that silver-containing hydrocolloid dressings were no better than standard alginate dressings in treating diabetic foot ulcers. A 2010 Cochrane review found insufficient evidence to determine if dressings containing silver increase or decrease infection or effect healing rates. Another 2010 review found some evidence that silver-impregnated dressings improve the short-term healing of wounds and ulcers.

The lead author of this paper is a speaker for one of the manufacturers of one of the silver dressings under study. A 2009 systematic review found that silver dressings improve both wound healing and quality of life when managing chronic non-healing wounds. Another 2009 review concluded that the evidence for silver-containing foam in chronic infected wounds is not clear, but found that silver-containing foam resulted in a greater reduction in wound size and more effective control of leakage and odor than non-silver dressings. A Cochrane review from 2008 found that, despite some potentially positive findings, most of the trials had methodological shortcomings and thus are of little use. The review also raised concerns about delays in time to wound healing and an increased number of dressing applications when silver sulfadiazine (SSD) is used for the full duration of the treatment. Another 2008 systematic review concluded that the evidence shows an overall positive effect of silver-releasing dressings in the management of infected chronic wounds, but expressed concern that the quality of the underlying trials was limited and potentially biased.

A number of wound dressings containing silver as an anti-bacterial have been cleared by the U.S. Food and Drug Administration (FDA).

Endotracheal Tubes

Limited evidence suggests that endotracheal breathing tubes coated with silver may reduce the incidence of ventilator associated pneumonia (VAP) and delay its onset, although no benefit is seen in the duration of intubation, the duration of stay in intensive care or the mortality rate. Concerns have been raised surrounding the unblinded nature of some of the studies. It is unknown if they are cost effective; and more high quality scientific trials are needed.

The U.S. Food and Drug Administration in 2007 cleared an endotracheal tube with a fine coat of silver to reduce the risk of ventilator-associated pneumonia.

Urinary Catheters

Tentative evidence supports a decreased risk of urinary tract infections when silver-alloy catheters are used. Their cost effectiveness is uncertain as of 2008.

Other Uses

Silver compounds are used in external preparations as antiseptics, including both silver nitrate and silver proteinate, which can be used in dilute solution as eyedrops to prevent conjunctivitis in newborn babies. Silver nitrate is also sometimes used in dermatology in solid stick form as a caustic ("lunar caustic") to treat certain skin conditions, such as corns and warts. Silver is also used in bone prostheses, reconstructive orthopedic surgery and cardiac devices.

Chlorhexidine-silver-sulfadiazine central venous catheters significantly reduce the incidence of catheter-related bloodstream infections (CR-BSI). Silver diamine fluoride is an effective intervention to reduce dental caries (tooth decay).

Silver acetate has been used as a potential aid to help stop smoking. A review of the literature in 2012, however, found no effect of silver acetate on smoking cessation at a six-month endpoint and if there is an effect it would be small.

Adverse Effects

In animals and humans, silver accumulates in the body. Chronic intake of silver products can result in an accumulation of silver or silver sulfide particles in the skin. These particles in the skin darken with exposure to sunlight, resulting in a blue or gray discolouration of the skin known as argyria. Localized argyria can occur as a result of topical use of silver-containing solutions, while generalized argyria results from the ingestion of such substances.

Argyria is generally irreversible, with the only practical method of minimizing its cosmetic disfigurement being to avoid the sun. Preliminary reports of treatment with laser therapy have been reported. These laser treatments are painful and general anesthesia is required. A similar laser treatment has been used to clear silver particles from the eye, a condition related to argyria called argyrosis. The Agency for Toxic Substances and Disease Registry (ATSDR) describes argyria as a "cosmetic problem".

While argyria is usually limited to skin discolouration, there are isolated reports of more serious neurologic, renal, or hepatic complications caused by ingesting colloidal silver.

Colloidal silver may interact with some prescription medications, reducing the absorption of some antibiotics and thyroxine among others.

Some people are allergic to silver, and the use of treatments and medical devices containing silver is contraindicated for such people. Although medical devices containing silver are widely used in hospitals, no thorough testing and standardization of these products has yet been undertaken.

Water Purification

Electrolytically-dissolved silver has been used as a water disinfecting agent, for example, the drinking water supplies of the Russian Mir orbital station and the International Space Station. Many modern hospitals filter hot water through copper-silver filters to defeat MRSA and legionella infections. The World Health Organisation includes silver in a colloidal state produced by electrolysis of silver electrodes in water, and colloidal silver in water filters as two of a number of water disinfection methods specified to provide safe drinking water in developing countries. Along these lines, a ceramic filtration system coated with silver particles has been created by Ron Rivera of Potters for Peace and used in developing countries for water disinfection (in this application the silver inhibits microbial growth on the filter substrate, to prevent clogging, and does not directly disinfect the filtered water).

Mechanism of Action

Silver and most silver compounds have an oligodynamic effect and are toxic for bacteria, algae, and fungi *in vitro*. Among the elements that have this effect, silver is the least toxic for humans. The antibacterial action of silver is dependent on the silver ion. The effectiveness of silver compounds as an antiseptic is based on the ability of the biologically active silver ion (Ag+) to irreversibly damage key enzyme systems in the cell membranes of pathogens. The antibacterial action of silver has long been known to be enhanced by the presence of an electric field. Applying an electric current across silver electrodes enhances antibiotic action at the anode, likely due to the release of silver into the bacterial culture. The antibacterial action of electrodes coated with silver nanostructures is greatly improved in the presence of an electric field.

Alternative Medicine

Since about 1990, there has been a resurgence of the promotion of colloidal silver as a dietary supplement or homeopathic remedy, marketed with claims of it being an essential mineral supplement, or that it can prevent or treat numerous diseases, such as cancer, diabetes, HIV/AIDS, herpes, and tuberculosis. No medical evidence supports the effectiveness of colloidal silver for any of these claimed indications. Silver is not an essential mineral in humans; there is no dietary requirement for silver, and no such thing as a silver "deficiency". There is no evidence that colloidal silver treats or prevents any medical condition, and it can cause serious and potentially irreversible side effects such as argyria. In August 1999, the U.S. FDA banned colloidal silver sellers from claiming any therapeutic or preventive value for the product, although silver-containing products continue to be promoted as dietary supplements in the U.S. under the looser regulatory standards applied to supplements. The FDA has issued numerous Warning Letters to Internet sites that have continued to promote colloidal silver as an antibiotic or for other medical purposes. Despite the efforts of the FDA, silver products remain widely available on the market today. A review of websites promoting nasal sprays containing colloidal silver suggested that information about silver-containing nasal sprays on the internet is misleading and inaccurate.

In 2002, the Australian Therapeutic Goods Administration (TGA) found there were no legitimate medical uses for colloidal silver and no evidence to support its marketing claims. The U.S. National Centre for Complementary and Alternative Medicine (NCCAM) warns that marketing claims about colloidal silver are scientifically unsupported, that the silver content of marketed supplements varies widely, and that colloidal silver products can have serious side effects such as argyria.

In 2009, the USFDA issued a "Consumer Advisory" warning about the potential adverse effects of colloidal silver, and said that "...there are no legally marketed prescription or over-the-counter (OTC) drugs containing silver that are taken by mouth." Quackwatch states that colloidal silver dietary supplements have not been found safe or effective for the treatment of any condition. *Consumer Reports* lists colloidal silver as a "supplement to avoid", describing it as "likely unsafe". The *Los Angeles Times* stated that "colloidal silver as a cure-all is a fraud with a long history, with quacks claiming it could cure cancer, AIDS, tuberculosis, diabetes and numerous other diseases."

History

Hippocrates in his writings discussed the use of silver in wound care. At the beginning of the twentieth century surgeons routinely used silver sutures to reduce the risk of infection. In the early 20th century, physicians used silver-containing eyedrops to treat ophthalmic problems, for various infections, and sometimes internally for diseases such as tropical sprue, epilepsy, gonorrhea, and the common cold. During World War I, soldiers used silver leaf to treat infected wounds. Prior to the introduction of modern antibiotics, colloidal silver was used as a germicide and disinfectant. With the development of modern antibiotics in the 1940s, the use of silver as an antimicrobial agent diminished. Silver sulfadiazine (SSD) is a compound containing silver and the antibiotic sodium sulfadiazine, which was developed in 1968.

Cost

The National Health Services in the UK spent about 25 million pounds on silver-containing dressing in 2006. Silver-containing dressings represent about 14% of the total dressings used and about 25% of the overall wound dressing costs. Concerns have been expressed about the potential environmental cost of manufactured silver nanomaterials in consumer applications being released into the environment, for example that they may pose a threat to benign soil organisms.

Nanoparticles to Deliver Antisense Oligonucleotides Aimed at Exon Skipping Therapies

Exon skipping is a method used to cause cells to "skip" over faulty or misaligned sections of genetic code, leading to a truncated but still functional protein despite the genetic mutation.

Exon skipping is used to restore the reading frame within a gene. Genes are the genetic instructions for creating a protein, and are composed of introns and exons. Exons are the sections of DNA that contain the instruction set for generating a protein; they are interspersed with non-coding regions called introns. The introns are later removed before the protein is made, leaving only the coding exon regions.

Splicing naturally occurs in pre-mRNA when introns are being removed to form mature-mRNA that consists solely of exons. Starting in the late 1990's, scientists realised they could take advantage of this naturally occurring cellular splicing to downplay genetic mutations into less harmful ones.

The mechanism behind exon skipping is a mutation specific antisense oligonucleotide (AONs). An antisense oligonucleotide is a synthesized short nucleic acid polymer, typically fifty or fewer base pairs in length that will bind to the mutation site in the pre-messenger RNA, to induce exon skipping. The AON binds to the mutated exon, so that when the gene is then translated from the mature mRNA, it is "skipped" over, thus restoring the disrupted reading frame. This allows for the generation of an internally deleted, but largely functional protein.

Some mutations require exon skipping at multiple sites, sometimes adjacent to one another, in order to restore the reading frame. Multiple exon skipping has successfully been carried out using a combination of AONs that target multiple exons.

Primary Application of Exon Skipping

Exon skipping is being heavily researched for the treatment of Duchenne muscular dystrophy (DMD), where the muscular protein dystrophin is prematurely truncated, which leads to a non-functioning protein. Successful treatment by way of exon skipping could lead to a mostly functional dystrophin protein, and create a phenotype similar to the less severe Becker muscular dystrophy (BMD).

In the case of Duchenne muscular dystrophy, the protein that becomes compromised is dystrophin. Dystrophin is the largest gene in the human body, composed of 79 exons. The dystrophin protein has two essential functional domains that flank a central rod domain consisting of repetitive and partially dispensable segments. Dystrophin's function is to maintain muscle fiber stability during contraction by linking the extra cellular matrix to the cytoskeleton. Mutations that disrupt the open reading frame within dystrophin create prematurely truncated proteins that are unable to perform their job. Such mutations lead to muscle fiber damage, replacement of muscle tissue by fat and fibrotic tissue, and premature death typically occurring in the early twenties of DMD patients. Comparatively, mutations that do not upset the open reading frame, lead to a dystrophin protein that is internally deleted and shorter than normal, but still partially functional. Such mutations are associated with the much milder Becker muscular dystrophy. Mildly affected BMD patients carrying deletions that involve over two thirds of the central rod domain have been described, suggesting that this domain is largely dispensable.

Dystrophin can maintain a large degree of functionality so long as the essential terminal domains are unaffected, and exon skipping only occurs within the central rod domain. Given these parameters, exon skipping can be used to restore an open reading frame by inducing a deletion of one or several exons within the central rod domain, and thus converting a DMD phenotype into a BMD phenotype.

The genetic mutation that leads to Becker muscular dystrophy is an in-frame deletion. This means the out of the 79 exons that code for dystrophin, one or several in the middle may be removed, without affecting the exons that follow the deletion. This allows for a shorter-than-normal dystrophin protein that maintains a degree of functionality. In Duchenne muscular dystrophy, the genetic mutation is out-of-frame. Out-of-frame mutations cause a premature stop in protein generation- the ribosome is unable to "read" the RNA past the point of initial error- leading to a severely shortened and completely non-functional dystrophin protein.

The goal of exon skipping is to manipulate the splicing pattern so that an out-of-frame mutation becomes an in-frame mutation, thus changing a severe DMD mutation into a less harmful in-frame BMD mutation.

Specific Applications of Exon Skipping to Duchenne Muscular Dystrophy

Genetic testing can be done, usually by a blood sample, to determine the precise nature and location of mutation to the dystrophin gene. Mutations within the dystrophin gene are known to cluster in certain areas, known as hot spot regions, primarily on exons 45-53 and to a lesser extent on exons 2-20. Antisense-mediated exon skipping to restore reading frames within dystrophin is currently being researched as a treatment for DMD, however; the anti-sense oligonucleotides (AONs) need to induce exon skipping are mutation specific. Fortunately, the majority of dystrophin mutations occur in the major hotspot region between exons 45-53, and the minor hot spot region ranging from exon 2-20. This allows for the creation of commonly needed AONs that would help a large number of DMD patients. By designing AON "cocktails" to induce the skipping of the eight exons in a hotspot region, up to 50% of DMD patients would be helped.

3

Oligonucleotide Synthesis

Oligonucleotide synthesis is the chemical synthesis of relatively short fragments of nucleic acids with defined chemical structure (sequence). The technique is extremely useful in current laboratory practice because it provides a rapid and inexpensive access to custom-made oligonucleotides of the desired sequence. Whereas enzymes synthesize DNA and RNA in a 5' to 3' direction, chemical oligonucleotide synthesis is carried out in the opposite, 3' to 5' direction. Currently, the process is implemented as solid-phase synthesis using phosphoramidite method and phosphoramidite building blocks derived from protected 2'-deoxynucleosides (dA, dC, dG, and T), ribonucleosides (A, C, G, and U), or chemically modified nucleosides, *e.g.* LNA, BNA.

To obtain the desired oligonucleotide, the building blocks are sequentially coupled to the growing oligonucleotide chain in the order required by the sequence of the product. The process has been fully automated since the late 1970s. Upon the completion of the chain assembly, the product is released from the solid phase to solution, deprotected, and collected. The occurrence of side reactions sets practical limits for the length of synthetic oligonucleotides (up to about 200 nucleotide residues) because the number of errors accumulates with the length of the oligonucleotide being synthesized. Products are often isolated by high-performance liquid chromatography (HPLC) to obtain the desired oligonucleotides in high purity. Typically, synthetic oligonucleotides are single-stranded DNA or RNA molecules around 15–25 bases in length.

Oligonucleotides find a variety of applications in molecular biology and medicine. They are most commonly used as antisense oligonucleotides, small interfering RNA, primers for DNA sequencing and amplification, probes for detecting complementary DNA or RNA via molecular hybridization, tools for the targeted introduction of mutations and restriction sites, and for the synthesis of artificial genes.

The evolution of oligonucleotide synthesis saw four major methods of the formation of internucleosidic linkages and has been reviewed in the literature in great detail.

Early Work and Contemporary H-Phosphonate Synthesis

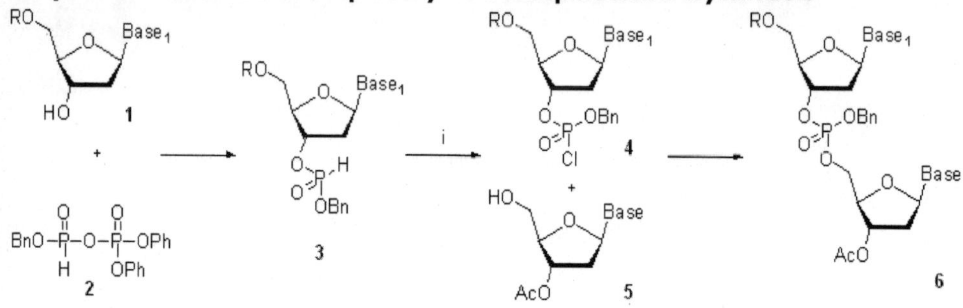

Figure: *Scheme. 1.* i: *N-Chlorosuccinimide; Bn = -CH_2Ph*

In the early 1950s, Alexander Todd's group pioneered H-phosphonate and phosphate triester methods of oligonucleotide synthesis. The reaction of compounds 1 and 2 to form H-phosphonate diester 3 is an H-phosphonate coupling in solution while that of compounds 4 and 5 to give 6 is a phosphotriester coupling.

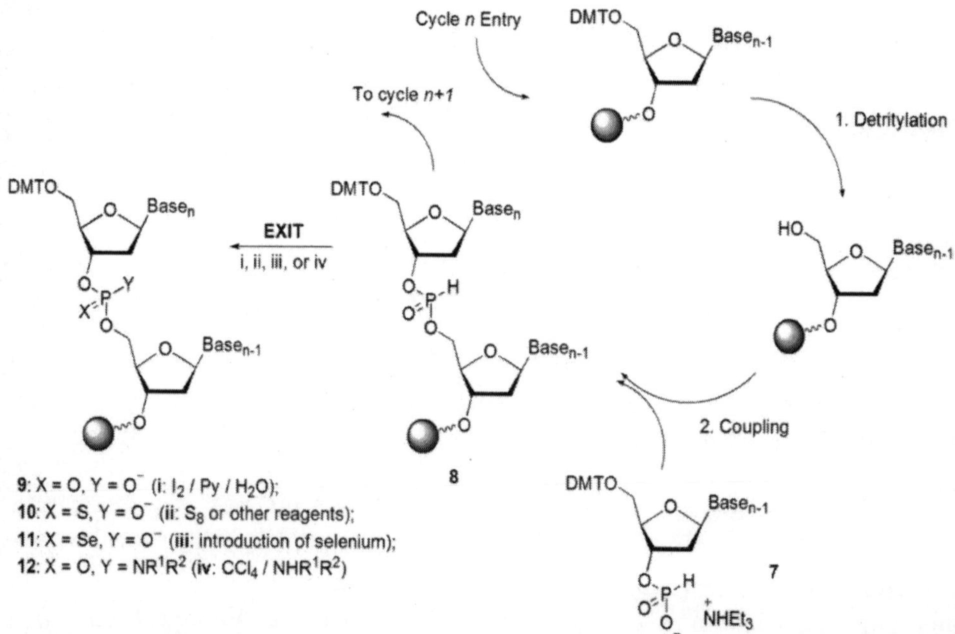

Figure: *Scheme 2. Synthesis of oligonucleotides by the H-Phosphonate Method*

Thirty years later, this work inspired, independently, two research groups to adopt the H-phosphonate chemistry to the solid-phase synthesis using nucleoside H-phosphonate monoesters 7 as building blocks and pivaloyl chloride, 2,4,6-triisopropylbenzenesulfonyl chloride (TPS-Cl), and other compounds as activators. The practical implementation of H-phosphonate method resulted in a very short and simple synthetic cycle consisting of only two steps, detritylation and coupling (Scheme 2). Oxidation of internucleosidic H-phosphonate diester linkages in 8 to phosphodiester linkages in 9 with a solution of iodine in aqueous pyridine is carried out at the end of the chain assembly rather than as a step in the synthetic cycle. If desired, the oxidation may be carried out under anhydrous conditions. Alternatively, 8 can be converted to phosphorothioate 10 or phosphoroselenoate 11 (X = Se), or oxidized by CCl_4 in the presence of primary or secondary amines to phosphoramidate analogs 12. The method is very convenient in that various types of phosphate modifications (phosphate/phosphorothioate/phosphoramidate) may be introduced to the same oligonucleotide for modulation of its properties.

Most often, H-phosphonate building blocks are protected at the 5'-hydroxy group and at the amino group of nucleic bases A, C, and G in the same manner as phosphoramidite building blocks. However, protection at the amino group is not mandatory.

Phosphodiester Synthesis

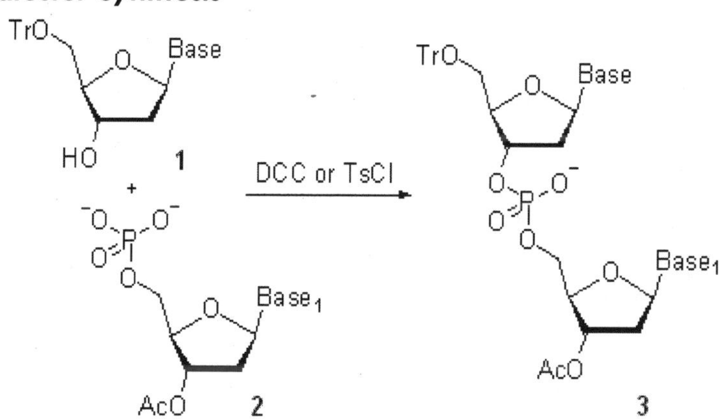

***Figure:** Scheme. 3 Oligonucleotide coupling by phosphodiester method; Tr = -CPh_3*

In the 1950s, Har Gobind Khorana and co-workers developed a phosphodiester method where 3'-*O*-acetylnucleoside-5'-*O*-phosphate 2 (Scheme 3) was activated with *N,N'*-dicyclohexylcarbodiimide (DCC) or 4-toluenesulfonyl chloride (Ts-Cl). The activated species were reacted with a 5'-*O*-protected nucleoside 1 to give a protected dinucleoside monophosphate 3. Upon the removal of 3'-*O*-acetyl group using base-catalyzed hydrolysis, further chain elongation was carried out. Following this methodology, sets of tri- and tetradeoxyribonucleotides were synthesized and were enzymatically converted to longer oligonucleotides, which allowed elucidation of the genetic code. The major limitation of the phosphodiester method consisted in the formation of pyrophosphate oligomers and oligonucleotides branched at the internucleosidic phosphate. The method seems to be a step back from the more selective chemistry described earlier; however, at that time, most phosphate-protecting groups available now had not yet been introduced. The lack of the convenient protection strategy necessitated taking a retreat to a slower and less selective chemistry to achieve the ultimate goal of the study.

Phosphotriester Synthesis

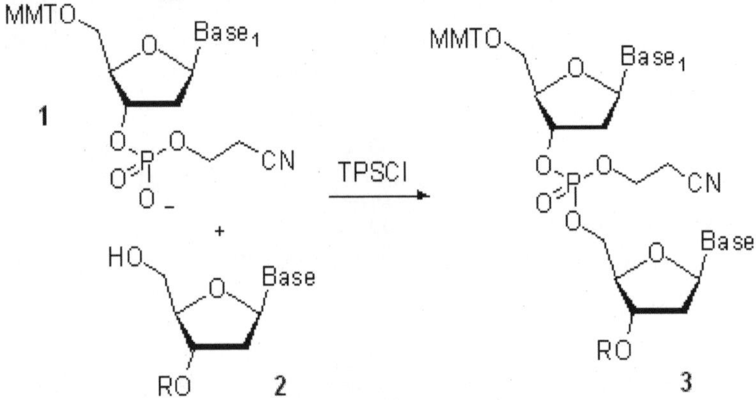

Figure: *Scheme 4. Oligonucleotide coupling by phosphotriester method; MMT = -$CPh_2(4\text{-}MeOC_6H_4)$.*

In the 1960s, groups led by R. Letsinger and C. Reese developed a phosphotriester approach. The defining difference from the phosphodiester approach was the protection of the phosphate moiety in the building block 1 (Scheme 4) and in the product 3 with 2-cyanoethyl group. This precluded the formation of oligonucleotides branched at the internucleosidic phosphate. The higher selectivity of the method allowed the use of more efficient coupling agents and catalysts, which dramatically reduced the length of the synthesis. The method, initially developed for the solution-phase synthesis, was also implemented on low-cross-linked "popcorn" polystyrene, and later on controlled pore glass, which initiated a massive research effort in solid-phase synthesis of oligonucleotides and eventually led to the automation of the oligonucleotide chain assembly.

Phosphite Triester Synthesis

In the 1970s, substantially more reactive P(III) derivatives of nucleosides, 3'-*O*-chlorophosphites, were successfully used for the formation of internucleosidic linkages. This led to the discovery of the phosphite triester methodology. The group led by M. Caruthers took the advantage of less aggressive and more selective 1*H*-tetrazolidophosphites and implemented the method on solid phase. Very shortly after, the workers from the same group further improved the method by using more stable nucleoside phosphoramidites as building blocks. The use of 2-cyanoethyl phosphite-protecting group in place of a less user-friendly methyl group led to the nucleoside phosphoramidites currently used in oligonucleotide synthesis. Many later improvements to the manufacturing of building blocks, oligonucleotide synthesizers, and synthetic protocols made the phosphoramidite chemistry a very reliable and expedite method of choice for the preparation of synthetic oligonucleotides.Synthesis by the phosphoramidite method

Nucleoside Phosphoramidites

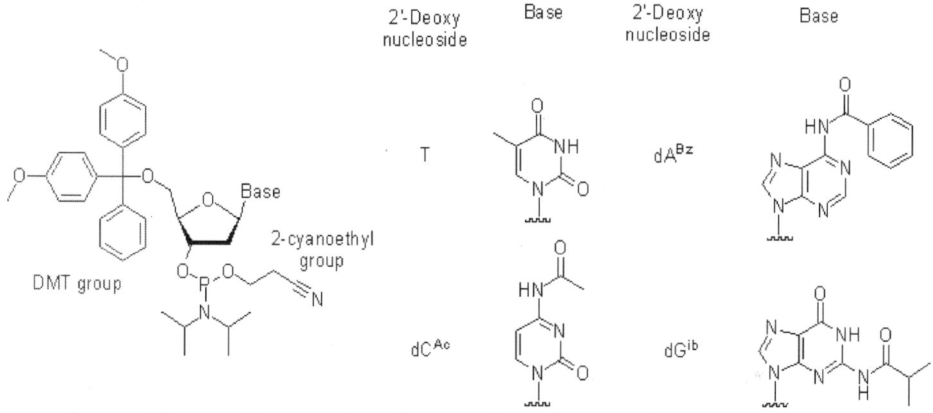

Figure: *Protected 2'-deoxynucleoside phosphoramidites.*

As mentioned above, the naturally occurring nucleotides (nucleoside-3'- or 5'-phosphates) and their phosphodiester analogs are insufficiently reactive to afford an expedite synthetic preparation of oligonucleotides in high yields. The selectivity and the rate of the formation of internucleosidic linkages is dramatically improved by using 3'-*O*-(*N*,*N*-diisopropyl phosphoramidite) derivatives of nucleosides (nucleoside phosphoramidites) that serve as building blocks in phosphite triester methodology. To prevent undesired side reactions, all other functional groups present in nucleosides have to be rendered unreactive (protected) by attaching protecting groups. Upon the completion of the oligonucleotide chain assembly, all the protecting groups are removed to yield the desired oligonucleotides. Below, the protecting groups currently used in commercially available and most common nucleoside phosphoramidite building blocks are briefly reviewed:

- The 5'-hydroxyl group is protected by an acid-labile DMT (4,4'-dimethoxytrityl) group.
- Thymine and uracil, nucleic bases of thymidine and uridine, respectively, do not have exocyclic amino groups and hence do not require any protection.

- Although the nucleic base of guanosine and 2'-deoxyguanosine does have an exocyclic amino group, its basicity is low to an extent that it does not react with phosphoramidites under the conditions of the coupling reaction.

  However, a phosphoramidite derived from the N2-unprotected 5'-*O*-DMT-2'-deoxyguanosine is poorly soluble in acetonitrile, the solvent commonly used in oligonucleotide synthesis. In contrast, the N2-protected versions of the same compound dissolve in acetonitrile well and hence are widely used. Nucleic bases adenine and cytosine bear the exocyclic amino groups reactive with the activated phosphoramidites under the conditions of the coupling reaction.

  By the use of additional steps in the synthetic cycle or alternative coupling agents and solvent systems, the oligonucleotide chain assembly may be carried out using dA and dC phosphoramidites with unprotected amino groups.

  However, these approaches currently remain in the research stage. In the routine oligonucleotide synthesis, exocyclic amino groups in nucleosides are kept permanently protected over the entire length of the oligonucleotide chain assembly.

The protection of the exocyclic amino groups has to be orthogonal to that of the 5'-hydroxy group because the latter is removed at the end of each synthetic cycle. The simplest to implement and hence the most widely accepted is the strategy where the exocyclic amino groups bear a base-labile protection. Most often, two protection schemes are used.

- In the first, the standard and more robust scheme (Figure), Bz (benzoyl) protection is used for A, dA, C, and dC, while G and dG are protected with isobutyryl group. More recently, Ac (acetyl) group is often used to protect C and dC as shown in Figure.
- In the second, mild protection scheme, A and dA are protected with isobutyryl or phenoxyacetyl groups (PAC). C and dC bear acetyl protection, and G and dG are protected with 4-isopropylphenoxyacetyl (*i*Pr-PAC) or dimethylformamidino (dmf) groups. Mild protecting groups are removed more readily than the standard protecting groups. However, the phosphoramidites bearing these groups are less stable when stored in solution.
- The phosphite group is protected by a base-labile 2-cyanoethyl group. Once a phosphoramidite has been coupled to the solid support-bound oligonucleotide and the phosphite moieties have been converted to the P(V) species, the presence of the phosphate protection is not mandatory for the successful conducting of further coupling reactions.

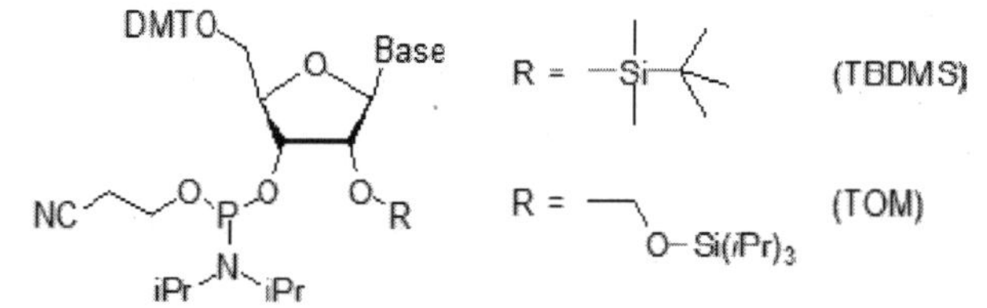

Figure: *2'-O-protected ribonucleoside phosphoramidites.*

- In RNA synthesis, the 2'-hydroxy group is protected with TBDMS (*t*-butyldimethylsilyl) group. or with TOM (tri-*iso*-propylsilyloxymethyl) group, both being removable by treatment with fluoride ion.
- The phosphite moiety also bears a diisopropylamino ($i\text{Pr}_2\text{N}$) group reactive under acidic conditions. Upon activation, the diisopropylamino group leaves to be substituted by the 5'-hydroxy group of the support-bound oligonucleotide.

Non-Nucleoside Phosphoramidites

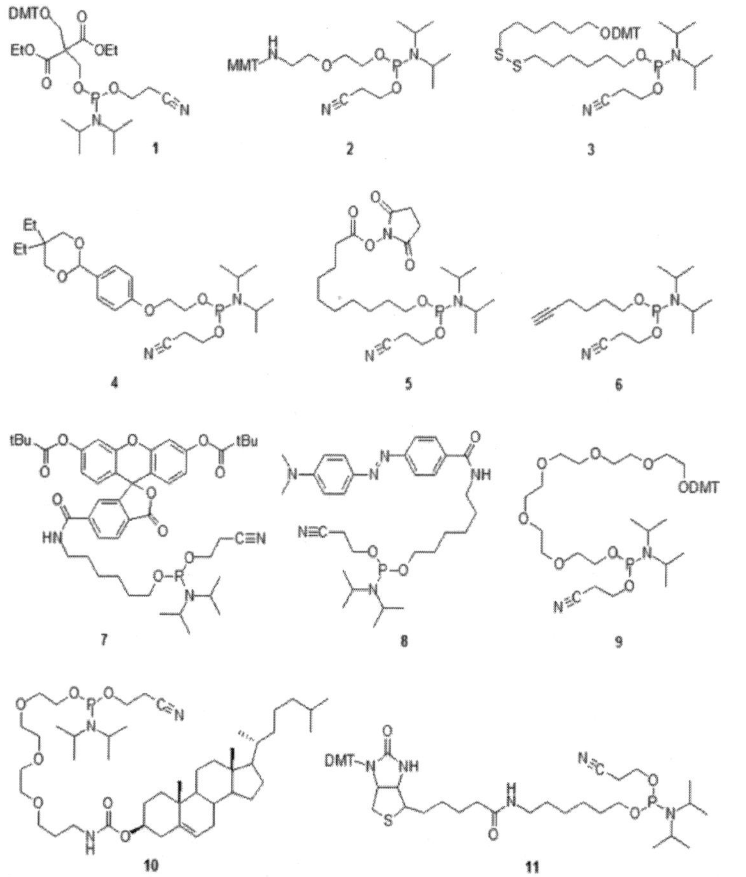

Figure: *Non-nucleoside phosphoramidites for 5'-modification of synthetic oligonucleotides. MMT = mono-methoxytrityl,(4-methoxyphenyl)diphenylmethyl.*

Non-nucleoside phosphoramidites are the phosphoramidite reagents designed to introduce various functionalities at the termini of synthetic oligonucleotides or between nucleotide residues in the middle of the sequence. In order to be introduced inside the sequence, a non-nucleosidic modifier has to possess at least two hydroxy groups, one of which is often protected with the DMT group while the other bears the reactive phosphoramidite moiety.

Non-nucleosidic phosphoramidites are used to introduce desired groups that are not available in natural nucleosides or that can be introduced more readily using simpler chemical designs. A very short selection of commercial phosphoramidite reagents is shown in Scheme for the demonstration of the available structural and functional diversity. These reagents serve for the attachment of 5'-terminal phosphate (1), NH_2 (2), SH (3), aldehydo (4), and carboxylic groups (5), CC triple bonds (6), non-radioactive labels and quenchers (exemplified by 6-FAM amidite 7 for the attachment of fluorescein and dabcyl amidite 8, respectively), hydrophilic and hydrophobic modifiers (exemplified by hexaethyleneglycol amidite 9 and cholesterol amidite 10, respectively), and biotin amidite 11.

Synthetic Cycle

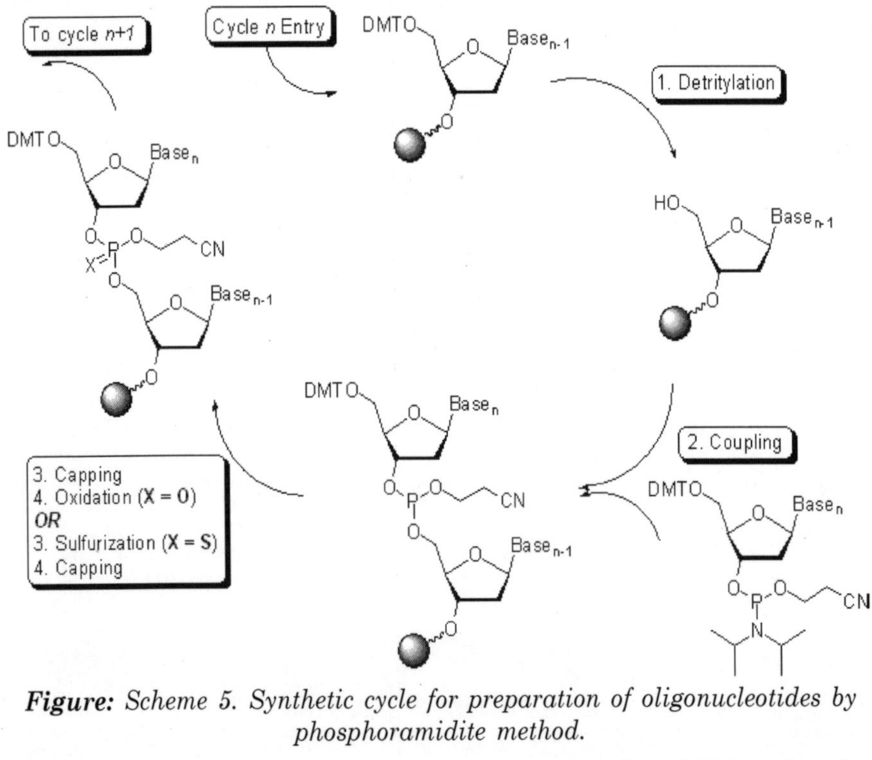

Figure: *Scheme 5. Synthetic cycle for preparation of oligonucleotides by phosphoramidite method.*

Oligonucleotide synthesis is carried out by a stepwise addition of nucleotide residues to the 5'-terminus of the growing chain until the desired sequence is assembled. Each addition is referred to as a synthetic cycle (Scheme 5) and consists of four chemical reactions:

Step 1: De-Blocking (Detritylation)

The DMT group is removed with a solution of an acid, such as 2% trichloroacetic acid (TCA) or 3% dichloroacetic acid (DCA), in an inert solvent (dichloromethane or toluene). The orange-coloured DMT cation formed is washed out; the step results in the solid support-bound oligonucleotide precursor bearing a free 5'-terminal hydroxyl group. It is worth remembering that conducting detritylation for an extended time or with stronger than recommended solutions of acids leads to depurination of solid support-bound oligonucleotide and thus reduces the yield of the desired full-length product.

Step 2: Coupling

A 0.02–0.2 M solution of nucleoside phosphoramidite (or a mixture of several phosphoramidites) in acetonitrile is activated by a 0.2–0.7 M solution of an acidic azole catalyst, 1*H*-tetrazole, 2-ethylthiotetrazole, 2-benzylthiotetrazole, 4,5-dicyanoimidazole, or a number of similar compounds. The mixing is usually very brief and occurs in fluid lines of oligonucleotide synthesizers while the components are being delivered to the reactors containing solid support. The activated phosphoramidite in 1.5 – 20-fold excess over the support-bound material is then brought in contact with the starting solid support (first coupling) or a support-bound oligonucleotide precursor (following couplings) whose 5'-hydroxy group reacts with the activated phosphoramidite moiety of the incoming nucleoside phosphoramidite to form a phosphite triester linkage. The coupling of 2'-deoxynucleoside phosphoramidites is very rapid and requires, on small scale, about 20 s for its completion. In contrast, sterically hindered 2'-*O*-protected ribonucleoside phosphoramidites require 5-15 min to be coupled in high yields. The reaction is also highly sensitive to the presence of water, particularly when dilute solutions of phosphoramidites are used, and is commonly carried out in anhydrous acetonitrile. Generally, the larger the scale of the synthesis, the lower the excess and the higher the concentration of the phosphoramidites is used. In contrast, the concentration of the activator is primarily determined by its solubility in acetonitrile and is irrespective of the scale of the synthesis. Upon the completion of the coupling, any unbound reagents and by-products are removed by washing.

Step 3: Capping

The capping step is performed by treating the solid support-bound material with a mixture of acetic anhydride and 1-methylimidazole or, less often, DMAP as catalysts and, in the phosphoramidite method, serves two purposes.

- After the completion of the coupling reaction, a small percentage of the solid support-bound 5'-OH groups (0.1 to 1%) remains unreacted and needs to be permanently blocked from further chain elongation to prevent the formation of oligonucleotides with an internal base deletion commonly referred to as (n-1) shortmers. The unreacted 5'-hydroxy groups are, to a large extent, acetylated by the capping mixture.

- It has also been reported that phosphoramidites activated with 1*H*-tetrazole react, to a small extent, with the O^6 position of guanosine. Upon oxidation with I_2 /water, this side product, possibly *via* O^6-N7 migration, undergoes depurination. The apurinic sites thus formed are readily cleaved in the course of the final deprotection of the oligonucleotide under the basic conditions to give two shorter oligonucleotides thus reducing the yield of the full-length product. The O^6 modifications are rapidly removed by treatment with the capping reagent as long as the capping step is performed *prior* to oxidation with I_2/water.
- The synthesis of oligonucleotide phosphorothioates (OPS) does not involve the oxidation with I_2/water, and, respectively, does not suffer from the side reaction described above. On the other hand, if the capping step is performed prior to sulfurization, the solid support may contain the residual acetic anhydride and N-methylimidazole left after the capping step. The capping mixture interferes with the sulfur transfer reaction, which results in the extensive formation of the phosphate triester internucleosidic linkages in place of the desired PS triesters. Therefore, for the synthesis of OPS, it is advisable to conduct the sulfurization step *prior* to the capping step.

Step 4: Oxidation

The newly formed tricoordinated phosphite triester linkage is not natural and is of limited stability under the conditions of oligonucleotide synthesis. The treatment of the support-bound material with iodine and water in the presence of a weak base (pyridine, lutidine, or collidine) oxidizes the phosphite triester into a tetracoordinated phosphate triester, a protected precursor of the naturally occurring phosphate diester internucleosidic linkage. Oxidation may be carried out under anhydrous conditions using tert-Butyl hydroperoxide or, more efficiently, (1S)-(+)-(10-camphorsulfonyl)-oxaziridine (CSO). The step of oxidation is substituted with a sulfurization step to obtain oligonucleotide phosphorothioates. In the latter case, the sulfurization step is best carried out prior to capping.

Solid Supports

In solid-phase synthesis, an oligonucleotide being assembled is covalently bound, via its 3'-terminal hydroxy group, to a solid support material and remains attached to it over the entire course of the chain assembly. The solid support is contained in columns whose dimensions depend on the scale of synthesis and may vary between 0.05 mL and several liters. The overwhelming majority of oligonucleotides are synthesized on small scale ranging from 40 nmol to 1 ìmol. More recently, high-throughput oligonucleotide synthesis where the solid support is contained in the wells of multi-well plates (most often, 96 or 384 wells per plate) became a method of choice for parallel synthesis of oligonucleotides on small scale. At the end of the chain assembly, the oligonucleotide is released from the solid support and is eluted from the column or the well.

Solid Support Material

In contrast to organic solid-phase synthesis and peptide synthesis, the synthesis of oligonucleotides proceeds best on non-swellable or low-swellable solid supports. The two most often used solid-phase materials are controlled pore glass (CPG) and macroporous polystyrene (MPPS).

- CPG is commonly defined by its pore size. In oligonucleotide chemistry, pore sizes of 500, 1000, 1500, 2000, and 3000 Å are used to allow the preparation of about 50, 80, 100, 150, and 200-mer oligonucleotides, respectively. To make native CPG suitable for further processing, the surface of the material is treated with (3-aminopropyl)triethoxysilane to give aminopropyl CPG. The aminopropyl arm may be further extended to result in long chain aminoalkyl (LCAA) CPG. The amino group is then used as an anchoring point for linkers suitable for oligonucleotide synthesis.
- MPPS suitable for oligonucleotide synthesis is a low-swellable, highly cross-linked polystyrene obtained by polymerization of divinylbenzene (min 60%), styrene, and 4-chloromethylstyrene in the presence of a porogeneous agent. The macroporous chloromethyl MPPS obtained is converted to aminomethyl MPPS.

Linker Chemistry

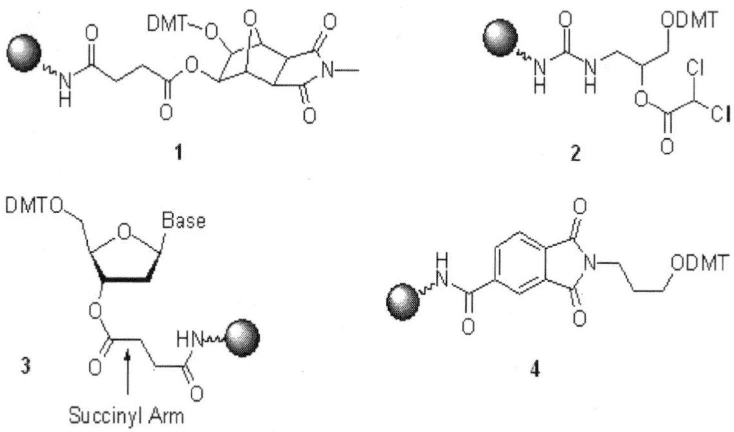

Figure: *Commercial solid supports for oligonucleotide synthesis.*

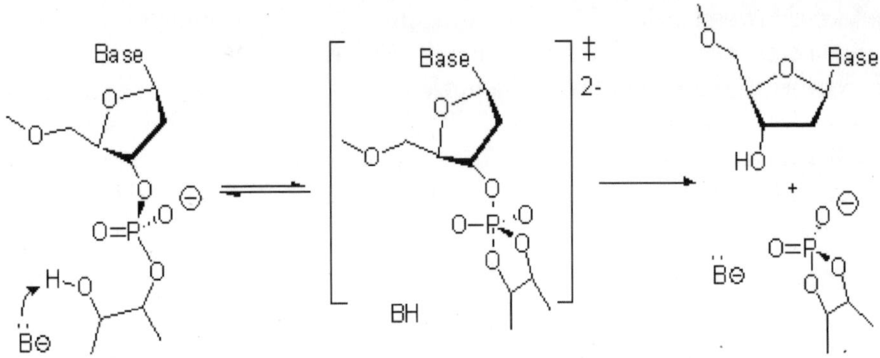

Figure: *Scheme 6. Mechanism of 3'-dephosphorylation of oligonucleotides assembled on universal solid supports.*

To make the solid support material suitable for oligonucleotide synthesis, non-nucleosidic linkers or nucleoside succinates are covalently attached to the reactive amino groups in aminopropyl CPG, LCAA CPG, or aminomethyl MPPS. The remaining unreacted amino groups are capped with acetic anhydride. Typically, three conceptually different groups of solid supports are used.

- *Universal supports*. In a more recent, more convenient, and more widely used method, the synthesis starts with the universal support where a non-nucleosidic linker is attached to the solid support material (compounds 1 and 2). A phosphoramidite respective to the 3'-terminal nucleoside residue is coupled to the universal solid support in the first synthetic cycle of oligonucleotide chain assembly using the standard protocols. The chain assembly is then continued until the completion, after which the solid support-bound oligonucleotide is deprotected. The characteristic feature of the universal solid supports is that the release of the oligonucleotides occurs by the hydrolytic cleavage of a P-O bond that attaches the 3'-*O* of the 3'-terminal nucleotide residue to the universal linker as shown in Scheme 6. The critical advantage of this approach is that the same solid support is used irrespectively of the sequence of the oligonucleotide to be synthesized. For the complete removal of the linker and the 3'-terminal phosphate from the assembled oligonucleotide, the solid support 1 and several similar solid supports require gaseous ammonia, aqueous ammonium hydroxide, aqueous methylamine, or their mixture and are commercially available. The solid support 2 requires a solution of ammonia in anhydrous methanol and is also commercially available.
- *Nucleosidic solid supports*. In a historically first and still popular approach, the 3'-hydroxy group of the 3'-terminal nucleoside residue is attached to the solid support via, most often, 3'-*O*-succinyl arm as in compound 3. The oligonucleotide chain assembly starts with the coupling

of a phosphoramidite building block respective to the nucleotide residue second from the 3'-terminus. The 3'-terminal hydroxy group in oligonucleotides synthesized on nucleosidic solid supports is deprotected under the conditions somewhat milder than those applicable for universal solid supports. However, the fact that a nucleosidic solid support has to be selected in a sequence-specific manner reduces the throughput of the entire synthetic process and increases the likelihood of human error.

- *Special solid supports* are used for the attachment of desired functional or reporter groups at the 3'-terminus of synthetic oligonucleotides. For example, the commercial solid support 4 allows the preparation of oligonucleotides bearing 3'-terminal 3-aminopropyl linker. Similarly to non-nucleosidic phosphoramidites, many other special solid supports designed for the attachment of reactive functional groups, non-radioactive reporter groups, and terminal modifiers (*e.c.* cholesterol or other hydrophobic tethers) and suited for various applications are commercially available. A more detailed information on various solid supports for oligonucleotide synthesis can be found in a recent review.

Oligonucleotide Phosphorothioates and Their Synthesis

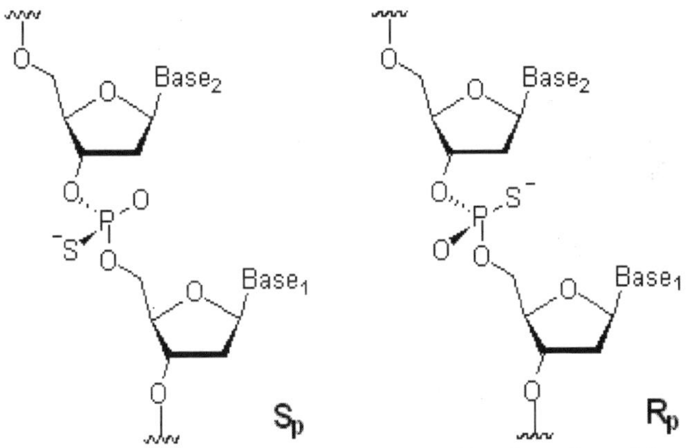

Figure: *S_p and R_p-diastereomeric internucleosidic phosphorothioate linkages.*

Oligonucleotide phosphorothioates (OPS) are modified oligonucleotides where one of the oxygen atoms in the phosphate moiety is replaced by sulfur. Only the phosphorothioates having sulfur at a non-bridging position as shown in figure are widely used and are available commercially. The replacement of the non-bridging oxygen with sulfur creates a new centre of chirality at phosphorus. In a simple case of a dinucleotide, this results in the formation of a diastereomeric pair of S_p- and R_p-dinucleoside monophosphorothioates whose structures are shown in Figure. In a *n*-mer oligonucleotide where all $(n-1)$ internucleosidic linkages are phosphorothioate linkages, the number of diastereomers *m* is calculated as $m = 2^{(n-1)}$. Being non-natural analogs of nucleic acids, OPS are substantially more stable towards hydrolysis by nucleases, the class of enzymes that destroy nucleic acids by breaking the bridging P-O bond of the phosphodiester moiety. This property determines the use of OPS as antisense oligonucleotides in *in vitro* and *in vivo* applications where the extensive exposure to nucleases is inevitable. Similarly, to improve the stability of siRNA, at least one phosphorothioate linkage is often introduced at the 3'-terminus of both sense and antisense strands. In chirally pure OPS, all-Sp diastereomers are more stable to enzymatic degradation than their all-Rp analogs. However, the preparation of chirally pure OPS remains a synthetic challenge. In laboratory practice, mixtures of diastereomers of OPS are commonly used.

Synthesis of OPS is very similar to that of natural oligonucleotides. The difference is that the oxidation step is replaced by sulfur transfer reaction (sulfurization) and that the capping step is performed after the sulfurization. Of many reported reagents capable of the efficient sulfur transfer, only three are commercially available:

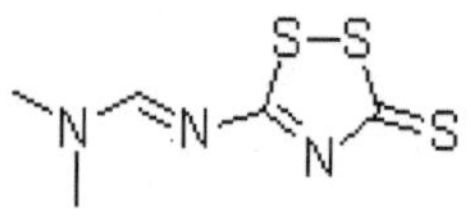

3

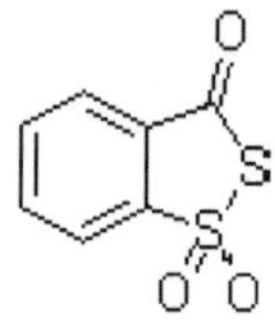

4

Figure: *Commercial sulfur transfer agents for oligonucleotide synthesis.*

- 3-(Dimethylaminomethylidene)amino)-3H-1,2,4-dithiazole-3-thione, DDTT (3) provides rapid kinetics of sulfurization and high stability in solution. The reagent is available from several sources.
- 3*H*-1,2-benzodithiol-3-one 1,1-dioxide (4) also known as Beaucage reagent displays a better solubility in acetonitrile and short reaction times. However, the reagent is of limited stability in solution and is less efficient in sulfurizing RNA linkages.
- *N,N,N'N'*-Tetraethylthiuram disulfide (TETD) is soluble in acetonitrile and is commercially available. However, the sulfurization reaction of an internucleosidic DNA linkage with TETD requires 15 min, which is more than 10 times as slow as that with compounds 3 and 4.

Automation

In the past, oligonucleotide synthesis was carried out manually in solution or on solid phase. The solid phase synthesis was implemented using, as containers for the solid phase, miniature glass columns similar in their shape to low-pressure chromatography columns or syringes equipped with porous filters. Currently, solid-phase oligonucleotide synthesis is carried out automatically using computer-controlled instruments (oligonucleotide synthesizers) and is technically implemented in column, multi-well plate, and array formats. The column format is best suited for research and large scale applications where a high-throughput is not required. Multi-well plate format is designed specifically for high-throughput synthesis on small scale to satisfy the growing demand of industry and academia for synthetic oligonucleotides. A number of oligonucleotide synthesizers for small scale synthesis and medium to large scale synthesis are available commercially.

Mid to Large Scale Oligonucleotide Synthesis - A History

Large scale oligonucleotide synthesizers were often developed by augmenting the capabilities of a preexisting instrument platform. One of the first mid scale synthesizers appeared in the late 1980's, manufactured by the Biosearch company in Novato, CA (The 8800). This platform was originally designed as a peptide synthesizer and made use of a fluidized bed reactor essential for accommodating the swelling characteristics of polystyrene supports used in the Merrifield methodology. Oligonucleotide synthesis involved the use of CPG (controlled pore glass) which is a rigid support and is more suited for column reactors as described above.

The scale of the 8800 was limited to the flow rate required to fluidize the support. Some novel reactor designs as well as higher than normal pressures enabled the 8800 to achieve scales that would prepare 1 mmole of oligonucleotide. In the mid 1990's several companies developed platforms that were based on semi-preparative and preparative liquid chromatographs. These system were well suited for a column reactor approach.

In most cases all that was required was to augment the number of fluids that could be delivered to the column. Oligo synthesis requires a minimum of 10 and liquid chromatographs usually accommodate 4. This was an easy design task and some semi-automatic strategies worked without any modifications to the preexisting LC equipment. PerSeptive Biosystems as well as Pharmacia (GE) were two of several companies that developed synthesizers out of liquid chromatographs. Genomic Technologies, Inc. was one of the few companies to develop a large scale oligonucleotide synthesizer that was, from the ground up, an oligonucleotide synthesizer.

The initial platform called the VLSS for very large scale synthesizer utilised large Pharmacia liquid chromatograph columns as reactors and could synthesize up to 75 millimoles of material. Many oligonucleotide synthesis factories designed and manufactured their own custom platforms and little is known due to the designs being proprietary. The VLSS design continued to be refined and is continued in the QMaster synthesizer which is a scaled down platform providing milligram to gram amounts of synthetic oligonucleotide.

Synthesis of Oligonucleotide Microarrays

One may visualize an oligonucleotide microarray as a miniature multi-well plate where physical dividers between the wells (plastic walls) are intentionally removed. With respect to the chemistry, synthesis of oligonucleotide microarrays is different from the conventional oligonucleotide synthesis in two respects:

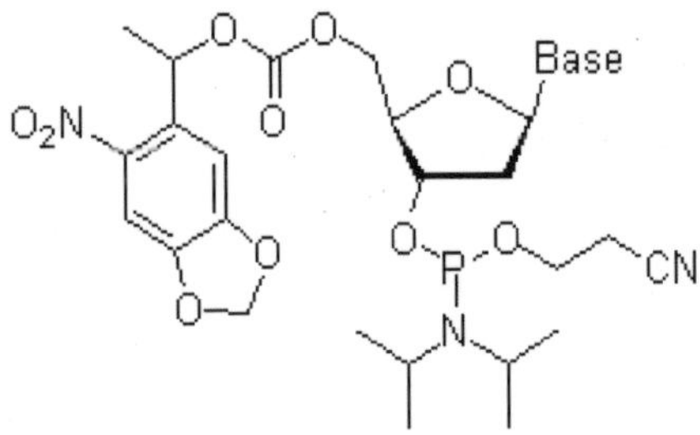

Figure: *5'-*O*-MeNPOC-protected nucleoside phosphoramidite.*

- Oligonucleotides remain permanently attached to the solid phase, which requires the use of linkers that are stable under the conditions of the final deprotection procedure.
- The absence of physical dividers between the sites occupied by individual oligonucleotides, a very limited space on the surface of the microarray (one oligonucleotide sequence occupies a square 25×25 ìm) and the requirement of high fidelity of oligonucleotide synthesis dictate the use of site-selective 5'-deprotection techniques. In one approach, the removal of the 5'-*O*-DMT group is effected by electrochemical generation of the acid at the required site(s). Another approach uses 5'-*O*-(á-methyl-6-nitropiperonyloxycarbonyl) (MeNPOC) protecting group, which can be removed by irradiation with UV light of 365 nm wavelength.

Post-Synthetic Processing

After the completion of the chain assembly, the solid support-bound oligonucleotide is fully protected:

- The 5'-terminal 5'-hydroxy group is protected with DMT group;
- The internucleosidic phosphate or phosphorothioate moieties are protected with 2-cyanoethyl groups;
- The exocyclic amino groups in all nucleic bases except for T and U are protected with acyl protecting groups.

To furnish a functional oligonucleotide, all the protecting groups have to be removed. The N-acyl base protection and the 2-cyanoethyl phosphate protection may be, and is often removed simultaneously by treatment with inorganic bases or amines. However, the applicability of this method is limited by the fact that the cleavage of 2-cyanoethyl phosphate protection gives rise to acrylonitrile as a side product. Under the strong basic conditions required for the removal of N-acyl protection, acrylonitrile is capable of alkylation of nucleic bases, primarily, at the N3-position of thymine and uracil residues to give the respective N3-(2-cyanoethyl) adducts *via* Michael reaction. The formation of these side products may be avoided by treating the solid support-bound oligonucleotides with solutions of bases in an organic solvent, for instance, with 50% triethylamine in acetonitrile or 10% diethylamine in acetonitrile. This treatment is strongly recommended for medium- and large scale preparations and is optional for syntheses on small scale where the concentration of acrylonitrile generated in the deprotection mixture is low.

Regardless of whether the phosphate protecting groups were removed first, the solid support-bound oligonucleotides are deprotected using one of the two general approaches.

(1) Most often, 5'-DMT group is removed at the end of the oligonucleotide chain assembly. The oligonucleotides are then released from the solid phase and deprotected (base and phosphate) by treatment with aqueous ammonium hydroxide, aqueous methylamine, their mixtures, gaseous ammonia or methylamine or, less commonly, solutions of other primary amines or alkalies at ambient or elevated temperature. This removes all remaining protection groups from 2'-deoxyoligonucleotides, resulting in a reaction mixture containing the desired product. If the oligonucleotide contains any 2'-*O*-protected ribonucleotide residues, the deprotection protocol includes the second step where the 2'-*O*-protecting silyl groups are removed by treatment with fluoride ion by various methods. The fully deprotected product is used as is, or the desired oligonucleotide can be purified by a number of methods. Most commonly, the crude product is desalted using ethanol precipitation, size exclusion chromatography, or reverse-phase HPLC. To eliminate unwanted truncation products, the oligonucleotides can be purified via polyacrylamide gel electrophoresis or anion-exchange HPLC followed by desalting.

(2) The second approach is only used when the intended method of purification is reverse-phase HPLC. In this case, the 5'-terminal DMT group that serves as a hydrophobic handle for purification is kept on at the end of the synthesis. The oligonucleotide is deprotected under basic conditions as described above and, upon evaporation, is purified by reverse-phase HPLC. The collected material is then detritylated under aqueous acidic conditions. On small scale (less than 0.01–0.02 mmol), the treatment with 80% aqueous acetic acid for 15–30 min at room temperature is often used followed by evaporation of the reaction mixture to dryness *in vacuo*. Finally, the product is desalted as described above.

- For some applications, additional reporter groups may be attached to an oligonucleotide using a variety of post-synthetic procedures.

Characterization

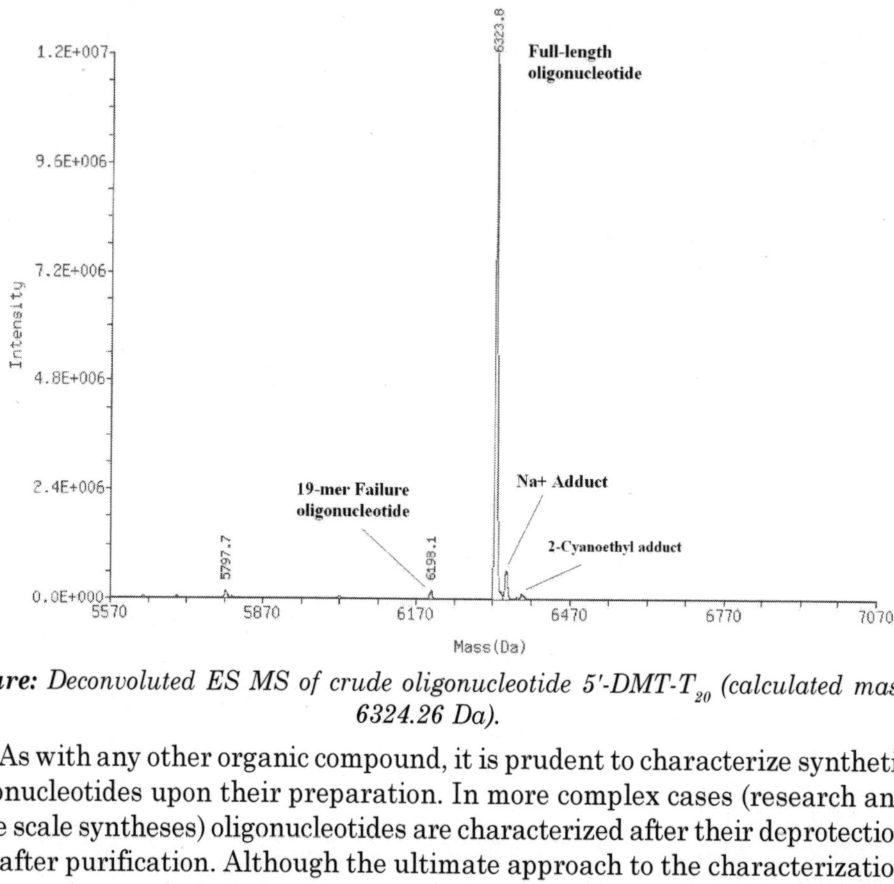

Figure: *Deconvoluted ES MS of crude oligonucleotide 5'-DMT-T_{20} (calculated mass 6324.26 Da).*

As with any other organic compound, it is prudent to characterize synthetic oligonucleotides upon their preparation. In more complex cases (research and large scale syntheses) oligonucleotides are characterized after their deprotection and after purification. Although the ultimate approach to the characterization is sequencing, a relatively inexpensive and routine procedure, the considerations of the cost reduction preclude its use in routine manufacturing of oligonucleotides. In day-by-day practice, it is sufficient to obtain the molecular mass of an oligonucleotide by recording its mass spectrum. Two methods are currently widely used for characterization of oligonucleotides: electrospray mass spectrometry (ES MS) and matrix-assisted laser desorption/ionization time-of-flight mass spectrometry (MALDI-TOF). To obtain informative spectra, it is very important to exchange all metal ions that might be present in the sample for ammonium or trialkylammonium [*e.c.* triethylammonium, $(C_2H_5)_3NH^+$] ions prior to submitting a sample to the analysis by either of the methods.

- In ES MS spectrum, a given oligonucleotide generates a set of ions that correspond to different ionization states of the compound. Thus, the oligonucleotide with molecular mass M generates ions with masses

(M – *n*H)/*n* where M is the molecular mass of the oligonucleotide in the form of a free acid (all negative charges of internucleosidic phosphodiester groups are neutralized with H^+), *n* is the ionization state, and H is the atomic mass of hydrogen (1 Da). Most useful for characterization are the ions with *n* ranging from 2 to 5. Software supplied with the more recently manufactured instruments is capable of performing a deconvolution procedure that is, it finds peaks of ions that belong to the same set and derives the molecular mass of the oligonucleotide.

- To obtain more detailed information on the impurity profile of oligonucleotides, liquid chromatography-mass spectrometry (LC-MS or HPLC-MS) or capillary electrophoresis mass spectrometry (CEMS) are used.

Antisense

Antisense RNA (asRNA) is a single-stranded RNA that is complementary to a messenger RNA (mRNA) strand transcribed within a cell. Some authors have used the term micRNA (mRNA-interfering complementary RNA) to refer to these RNAs but it is not widely used.

Antisense RNA may be introduced into a cell to inhibit translation of a complementary mRNA by base pairing to it and physically obstructing the translation machinery. This effect is therefore stoichiometric. An example of naturally occurring mRNA antisense mechanism is the hok/sok system of the *E. coli* R1 plasmid. Antisense RNA has long been thought of as a promising technique for disease therapy; the only such case to have reached the market is the drug fomivirsen. One commentator has characterized antisense RNA as one of "dozens of technologies that are gorgeous in concept, but exasperating in [commercialization]". Generally, antisense RNA still lack effective design, biological activity, and efficient route of administration.

Historically, the effects of antisense RNA have often been confused with the effects of RNA interference (RNAi), a related process in which *double-stranded* RNA fragments called small interfering RNAs trigger catalytically mediated gene silencing, most typically by targeting the RNA-induced silencing complex (RISC) to bind to and degrade the mRNA. Attempts to genetically engineer transgenic plants to express antisense RNA instead activate the RNAi pathway, although the processes result in differing magnitudes of the same downstream effect; gene silencing. Well-known examples include the Flavr Savr tomato and two cultivars of ringspot-resistant papaya.

Transcription of longer *cis*-antisense transcripts is a common phenomenon in the mammalian transcriptome. Although the function of some cases have been described, such as the Zeb2/Sip1 antisense RNA, no general function has been elucidated. In the case of Zeb2/Sip1, the antisense noncoding RNA is opposite the 5' splice site of an intron in the 5'UTR of the Zeb2 mRNA. Expression of the antisense ncRNA prevents splicing of an intron that contains a ribosome entry site necessary for efficient expression of the Zeb2 protein. Transcription of long antisense ncRNAs is often concordant with the associated protein-coding gene, but more detailed studies have revealed that the relative expression patterns of the mRNA and antisense ncRNA are complex.

Cis-Natural Antisense Transcript

Natural antisense transcripts (NATs) are a group of RNAs encoded within a cell that have transcript complementarity to other RNA transcripts. They have been identified in multiple eukaryotes, including humans, mice, yeast and *Arabidopsis thaliana*. This class of RNAs includes both protein-coding and non-coding RNAs. Current evidence has suggested a variety of regulatory roles for NATs, such as RNA interference (RNAi), alternative splicing, genomic imprinting, and X-chromosome inactivation. NATs are broadly grouped into two categories based on whether they act in cis or in trans. Trans-NATs are transcribed from a different location than their targets and usually have complementarity to multiple transcripts with some mismatches. MicroRNAs (miRNA) are an example of trans-NATs that can target multiple transcripts with a few mismatches. Cis-natural antisense transcripts (cis-NATs) on the other hand are transcribed from the same genomic locus as their target but from the opposite DNA strand and form perfect pairs.

Orientation of cis-NATs

Cis-NATs have a variety of orientations and differing lengths of overlap between pairs. There have been five identified orientations for cis-NATs to date. The most common orientation is head-to-head, where the 5' ends of both transcripts align together. This orientation would result in the greatest knockdown of gene expression if transcriptional collision is the reason for transcript inhibition. There are however some studies that have suggested that tail-to-tail orientations are the most common NAT pairs. Others such as tail to tail, overlapping, nearby head-to- head, and nearby tail-to-tail are less frequently encountered. Completely overlapping NATs involve the antisense gene being located completely over top of each other. Nearby head-to-head and tail-to-tail orientations are physically discrete from each other but are located very close to each other. Current evidence suggests that there is an overrepresentation of NAT pairs in genes that have catalytic activity. There may be something about these genes in particular that makes them more prone to this type of regulation.

Identification Approach

Identification of NATs in whole genomes is possible due to the large collection of sequence data available from multiple organisms. *In silico* methods for detecting NATs suffer from several shortcomings depending on the source of sequence information. Studies that use mRNA have sequences whose orientations are known, but the amount of mRNA sequence information available is small. Predicted gene models using algorithms trained to look for genes gives an increased coverage of the genome at the cost of confidence in the identified gene. Another resource is the extensive expressed sequence tag (EST) libraries but these small sequences must first be assigned an orientation before useful information can be extracted from them. Some studies have utilised special sequence information in the ESTs such as the poly(A) signal, poly(A) tail, and splicing sites to both filter the ESTs and to give them the correct transcriptional orientation. Combinations of the different sequence sources attempts to maximize coverage as well as maintain integrity in the data.

Pairs of NATs are identified when they form overlapping clusters. There is variability in the cut-off values used in different studies but generally ~20 nucleotides of sequence overlap is considered the minimum for transcripts to be considered and overlapping cluster. Also, transcripts must map to only one other mRNA molecule in order for it to be considered a NAT pair. Currently there are a variety of web and software resources that can be used to look for antisense pairs. The NATsdb or Natural Antisense Transcript database is a rich tool for searching for antisense pairs from multiple organisms.

Mechanisms

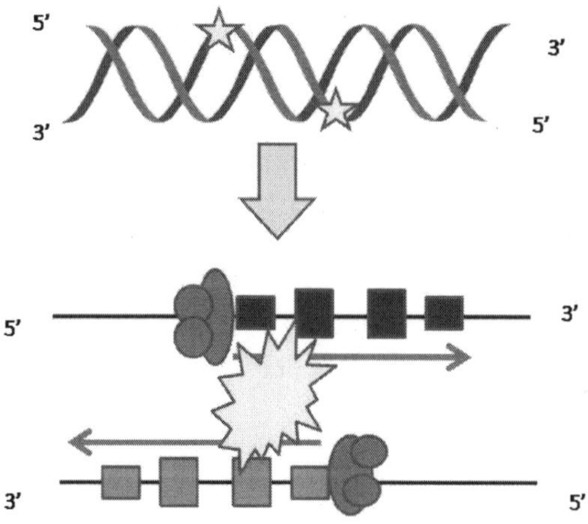

Figure: *Transcription collision model for expression inhibition*

Molecular mechanisms behind the regulatory role of cis-NATs are not currently well understood. Three models have been proposed to explain the regulatory effects that cis-NATs have on gene expression. The first model attributes that base pairing between the cis-NAT and its complementary transcript result in a knockdown of mRNA expression. The assumption of this model is that there will be a precise alignment of at least 6 base pairs between the cis-NAT pair to make double stranded RNA. Epigenetic modifications like DNA methylation and post-translational modification of core histones form the basis of the second model. Although it is not yet clearly understood, it is thought that the reverse transcript guides methylation complexes and/or histone-modifying complexes to the promoter regions of the sense transcript and cause an inhibition of expression from the gene. Currently it is not known what attributes of cis-NATs are crucial for the epigenetic model of regulation. The final proposed model that has gained favour due to recent experimental evidence is the transcriptional collision model. During the process of transcription of cis-NATs, the transcriptional complexes assemble in the promoter regions of the gene. RNA polymerases will then begin transcribing the gene at the transcription initiation site laying down nucleotides in a 5' to 3' direction. In the areas of overlap between the cis-NATs the RNA polymerases will collide and stop at the crash site. Transcription is inhibited because RNA polymerases prematurely stop and their incomplete transcripts get degraded.

Importance

Regulation of many biological processes such as development, metabolism and many others requires a careful co-ordination between many different genes; this is usually referred to as a gene regulatory network. A flurry of interest in gene regulatory networks has been sparked by the advent of sequenced genomes of multiple organisms. The next step is to use this information to figure out how genes work together and not just in isolation. During the processes of mammalian development, there is an inactivation of the extra X-chromosome in females. It has been shown that a NAT pair called Xist and Tsix are involved in the hypermethylation of the chromosome. As much as 20-30% of mammalian genes have been shown to be the targets of miRNAs, which highlights the importance of these molecules as regulators across a wide number of genes. Evolutionary reasons for utilising RNA for regulation of genes may be that it is less costly and faster than synthesizing proteins not needed by the cell. This could have had a selective advantage for early eukaryotes with this type of transcriptional regulation.

Cis-NATs and Disease

Antisense transcription might contribute to disease through chromosomal changes that result in the production of aberrant antisense transcripts. A documented case of cis-NATs being involved in human disease comes from an inherited form of á-thalassemia where there is silencing of the hemoglobin á-2 gene through the action of a cis-NAT. It is thought that in malignant cancer cells with activated transposable elements creates a large amount of transcriptional noise. It is likely that aberrant antisense RNA transcripts resulting from this transcriptional noise may cause stochastic methylation of CpG islands associated with oncogenes and tumor suppressor genes. This inhibition would further progress the malignancy of the cells since they lose key regulator genes. By looking at upregulated antisense transcripts in tumor cells, researchers and look for more candidate tumor suppressor genes. Also, aberrant cis-NATs have been implicated in neurological diseases such as Parkinson's disease.

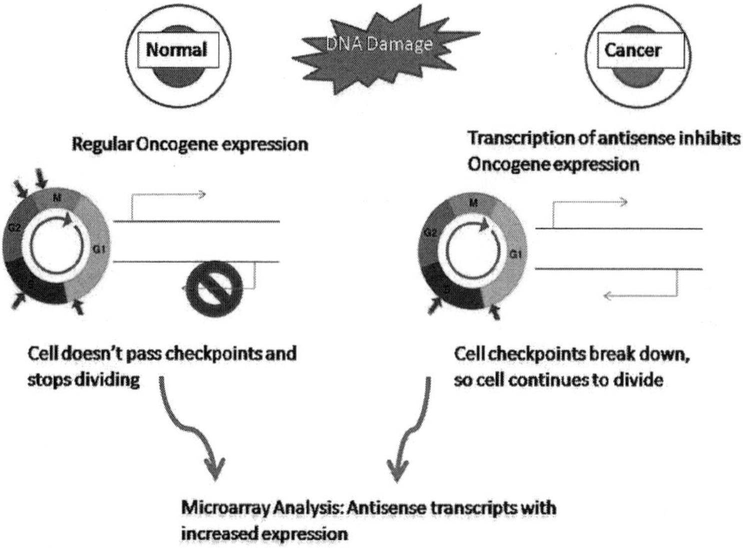

Figure: *Aberrant transcription of antisense transcripts can result in inhibition of oncogenes and allow cell to continue past cell cycle check points. Putative new oncogenes and tumor suppressor genes can be found by looking for upregulated antisense transcripts in cancer cells.*

Antisense Therapy

Antisense therapy is a form of treatment for genetic disorders or infections. When the genetic sequence of a particular gene is known to be causative of a particular disease, it is possible to synthesize a strand of nucleic acid (DNA, RNA or a chemical analogue) that will bind to the messenger RNA (mRNA) produced by that gene and inactivate it, effectively turning that gene "off". This is because mRNA has to be single stranded for it to be translated. Alternatively, the strand might be targeted to bind a splicing site on pre-mRNA and modify the exon content of an mRNA.

This synthesized nucleic acid is termed an "anti-sense" oligonucleotide because its base sequence is complementary to the gene's messenger RNA (mRNA), which is called the "sense" sequence (so that a sense segment of mRNA " 5'-AAGGUC-3' " would be blocked by the anti-sense mRNA segment " 3'-UUCCAG-5' ").

Antisense drugs are being researched to treat a variety of diseases such as cancers (including lung cancer, colorectal carcinoma, pancreatic carcinoma, malignant glioma and malignant melanoma), diabetes, Amyotrophic lateral sclerosis (ALS), Duchenne muscular dystrophy and diseases such as asthma, arthritis and pouchitis with an inflammatory component. As of 2014 two antisense drugs have been approved by the U.S. Food and Drug Administration (FDA), fomivirsen (marketed as Vitravene) as a treatment for cytomegalovirus retinitis and mipomersen (marketed as Kynamro) for homozygous familial hypercholesterolemia.

Example Antisense Therapies

As of 2012, some 40 antisense oligonucleotides and siRNAs were in clinical trials, including over 20 in advanced clinical trials (Phase II or III).

Cytomegalovirus Retinitis

Fomivirsen (marketed as Vitravene), was approved by the U.S. FDA in Aug 1998 as a treatment for cytomegalovirus retinitis.

Hemorrhagic Fever Viruses

In early 2006, scientists studying the Ebola hemorrhagic fever virus at USAMRIID announced a 75% recovery rate after infecting four rhesus monkeys and then treating them with an antisense Morpholino drug developed by Sarepta Therapeutics (formerly named AVI BioPharma), a U.S. biotechnology firm. The usual mortality rate for monkeys infected with Ebola virus is 100%. In late 2008, AVI BioPharma successfully filed Investigational New Drug (IND) applications with the FDA for its two lead products for Marburg and Ebola viruses. These drugs, AVI-6002 and AVI-6003 are novel analogs based on AVI's PMO antisense chemistry in which anti-viral potency is enhanced by the addition of positively-charged components to the morpholino oligomer chain. Preclinical results of AVI-6002 and AVI-6003 demonstrated reproducible and high rates of survival in non-human primates challenged with a lethal infection of the Ebola and Marburg viruses, respectively.

Cancer

Also in 2006, German physicians reported on a dose-escalation study for the compound AP 12009 (a phosphorothioate antisense oligodeoxynucleotide specific for the mRNA of human transforming growth factor TGF-beta2) in patients with high grade gliomas. At the time of the report, the median overall survival had not been obtained and the authors hinted at a potential cure.

HIV/AIDS

Starting in 2004, researchers in the US have been conducting research on using antisense technology to combat HIV.

In February 2010 researchers reported success in reducing HIV viral load using patient T-cells which had been harvested, modified with an RNA antisense strand to the HIV viral envelope protein, and re-infused into the patient during a planned lapse in retroviral drug therapy.

Familial Hypercholesterolemia

In January 2013 mipomersen (marketed as Kynamro) was approved by the FDA for the treatment of homozygous familial hypercholesterolemia.

Delivery

Because nucleases that cleave the phosphodiester linkage in DNA are expressed in most every cell, unmodified DNA molecules are generally degraded before their reach their targets, and so antisense drug candidate molecules are generally modified during the drug discovery phase of their development. Additionally, most targets of antisense are located inside cells, and getting nucleic acids across cell membranes is also difficult.

Therefore most clinical candidates have modified DNA "backbones", or the nucleobase or sugar moieties of the nucleotides are altered. Additionally, other molecules may be conjugated to antisense molecules in order to improve their ability to target certain cells or to cross barriers like cell membranes or the blood brain barrier.

Antisense mRNA

Antisense RNA (asRNA) is a single-stranded RNA that is complementary to a messenger RNA (mRNA) strand transcribed within a cell. Some authors have used the term micRNA (mRNA-interfering complementary RNA) to refer to these RNAs but it is not widely used.

Antisense RNA may be introduced into a cell to inhibit translation of a complementary mRNA by base pairing to it and physically obstructing the translation machinery. This effect is therefore stoichiometric. An example of naturally occurring mRNA antisense mechanism is the hok/sok system of the *E. coli* R1 plasmid. Antisense RNA has long been thought of as a promising technique for disease therapy; the only such case to have reached the market is the drug fomivirsen. One commentator has characterized antisense RNA as one of "dozens of technologies that are gorgeous in concept, but exasperating in [commercialization]". Generally, antisense RNA still lack effective design, biological activity, and efficient route of administration.

Historically, the effects of antisense RNA have often been confused with the effects of RNA interference (RNAi), a related process in which *double-stranded* RNA fragments called small interfering RNAs trigger catalytically mediated gene silencing, most typically by targeting the RNA-induced silencing complex (RISC) to bind to and degrade the mRNA. Attempts to genetically engineer transgenic plants to express antisense RNA instead activate the RNAi pathway, although the processes result in differing magnitudes of the same downstream effect; gene silencing. Well-known examples include the Flavr Savr tomato and two cultivars of ringspot-resistant papaya.

Transcription of longer *cis*-antisense transcripts is a common phenomenon in the mammalian transcriptome. Although the function of some cases have been described, such as the Zeb2/Sip1 antisense RNA, no general function has been elucidated. In the case of Zeb2/Sip1, the antisense noncoding RNA is opposite the 5' splice site of an intron in the 5'UTR of the Zeb2 mRNA. Expression of the antisense ncRNA prevents splicing of an intron that contains a ribosome entry site necessary for efficient expression of the Zeb2 protein. Transcription of long antisense ncRNAs is often concordant with the associated protein-coding gene, but more detailed studies have revealed that the relative expression patterns of the mRNA and antisense ncRNA are complex.

Morpholino

Morpholino, in molecular biology, describes a molecule in a particular structural family that is used to modify gene expression. Morpholino oligomers (oligos) are a class of antisense, a technology used to block access of other molecules to specific sequences within nucleic acids. Morpholinos block small (~25 base) regions of the base-pairing surfaces of ribonucleic acid (RNA). Most Morpholinos are used as research tools for reverse genetics by knocking down gene function. This article discusses only the Morpholino antisense oligos, which are nucleic acid analogs. The word "morpholino" can occur in other chemical names, referring to chemicals containing a six-membered morpholine ring. To help avoid confusion with other morpholine-containing molecules, when describing oligos "Morpholino" is often capitalized as a trade name, though this usage is not consistent throughout the scientific literature. Morpholinos are sometimes referred to as "PMO" (phosphorodiamidate morpholino oligo).

Gene knockdown is achieved by preventing cells from making a targeted protein. Morpholinos can also modify the splicing of pre-mRNA. Knocking down gene expression is a powerful method for learning about the function of a particular protein; in similar manner, causing a specific exon to be spliced out of a protein can help to determine the function of the protein moiety encoded by that exon or can sometimes knock down the protein activity altogether. These molecules have been applied to studies in several model organisms, including mice, zebrafish, frogs, and sea urchins.

Morpholinos are also in development as pharmaceutical therapeutics targeted against pathogenic organisms such as bacteria or viruses and for amelioration of genetic diseases. These synthetic oligos were conceived by James E. Summerton (Gene Tools, LLC) and developed in collaboration with Dwight D. Weller (formerly Antivirals Inc, renamed AVI BioPharma Inc., renamed Sarepta_Therapeutics Inc.).

Structure

Morpholinos are synthetic molecules that are the product of a redesign of natural nucleic acid structure. Usually 25 bases in length, they bind to complementary sequences of RNA by standard nucleic acid base-pairing. In terms of structure, the difference between Morpholinos and DNA is that, while Morpholinos have standard nucleic acid bases, those bases are bound to morpholine rings instead of deoxyribose rings and linked through phosphorodiamidate groups instead of phosphates. This may be easiest to visualize by referring to the first figure and comparing the structures of the two strands depicted there, one of RNA and the other of a Morpholino. Replacement of anionic phosphates with the uncharged phosphorodiamidate groups eliminates ionization in the usual physiological pH range, so Morpholinos in organisms or cells are uncharged molecules. The entire backbone of a Morpholino is made from these modified subunits.

Function

Morpholinos do not degrade their target RNA molecules, unlike many antisense structural types (e.g., phosphorothioates, siRNA). Instead, Morpholinos act by "steric blocking", binding to a target sequence within an RNA and simply getting in the way of molecules that might otherwise interact with the RNA. Morpholino oligos are often used to investigate the role of a specific mRNA transcript in an embryo. Developmental biologists inject Morpholino oligos into eggs or embryos of zebrafish, African clawed frog (Xenopus), sea urchin, and killifish (*F. heteroclitus*) producing morphant embryos, or electroporate Morpholinos into chick embryos at later stages of development. With appropriate cytosolic delivery systems, Morpholinos are effective in cell culture. Vivo-Morpholinos, in which the oligo is covalently linked to a delivery dendrimer, enter cells when administered systemically in adult animals. Morpholinos are being developed as pharmaceuticals under the name "NeuGenes" by Sarepta Therapeutics Inc. and are currently being tested in clinical trials for treatment of Duchenne muscular dystrophy in humans.

Normal Gene Expression in Eukaryotes

In eukaryotic organisms, pre-mRNA is transcribed in the nucleus, introns are spliced out, then the mature mRNA is exported from the nucleus to the cytoplasm. The small subunit of the ribosome usually starts by binding to one end of the mRNA and is joined there by various other eukaryotic initiation factors, forming the initiation complex. The initiation complex scans along the mRNA strand until it reaches a start codon, and then the large subunit of the ribosome attaches to the small subunit and translation of a protein begins. This entire process is referred to as gene expression; it is the process by which the information in a gene, encoded as a sequence of bases in DNA, is converted into the structure of a protein. A Morpholino can modify splicing or block translation, depending on the Morpholino's base sequence.

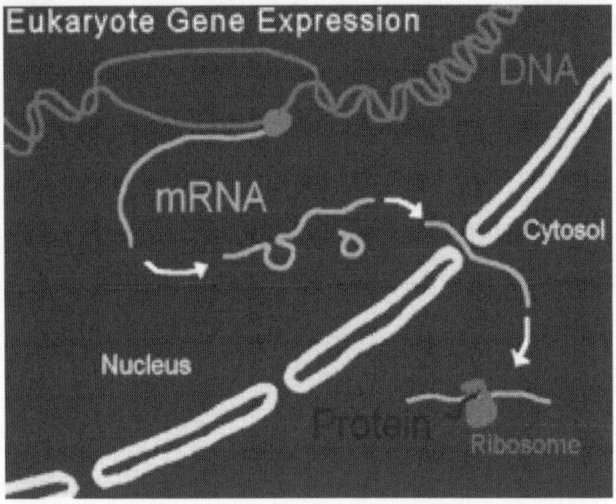

Figure: *Eukaryotic gene expression without intervention by a Morpholino*

Blocking Translation

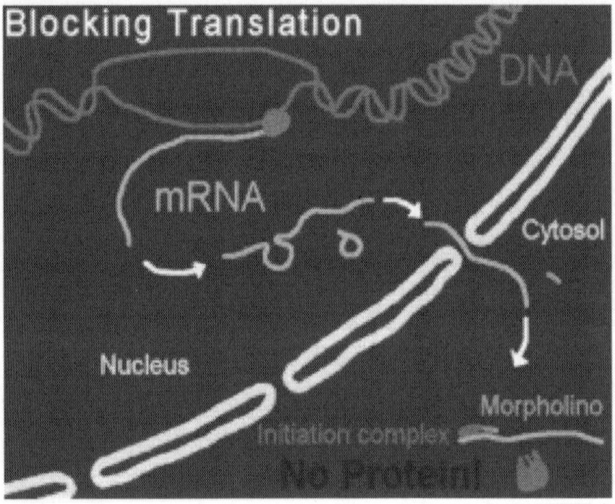

Figure: *Translation blocked by a Morpholino oligo*

Bound to the 5'-untranslated region of messenger RNA (mRNA), Morpholinos can interfere with progression of the ribosomal initiation complex from the 5' cap to the start codon. This prevents translation of the coding region of the targeted transcript (called "knocking down" gene expression). This is useful experimentally when an investigator wishes to know the function of a particular protein; Morpholinos provide a convenient means of knocking down expression of the protein and learning how that knockdown changes the cells or organism.

Some Morpholinos knock down expression so effectively that, after degradation of preexisting proteins, the targeted proteins become undetectable by Western blot (e.g., figure 1 in:).

Modifying Pre-mRNA Splicing

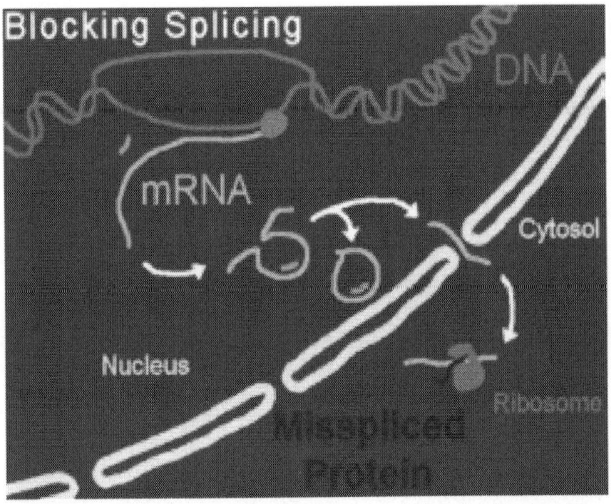

Figure: *Splicing blocked by a Morpholino oligo*

Morpholinos can interfere with pre-mRNA processing steps either by preventing splice-directing small nuclear ribonucleoproteins (snRNP) complexes from binding to their targets at the borders of introns on a strand of pre-mRNA, or by blocking the nucleophilic adenine base and preventing it from forming the splice lariat structure, or by interfering with the binding of splice regulatory proteins such as splice silencers and splice enhancers.

Preventing the binding of snRNP U1 (at the donor site) or U2/U5 (at the polypyrimidine moiety and acceptor site) can cause modified splicing, commonly excluding exons from the mature mRNA. Targeting some splice targets results in intron inclusions, while activation of cryptic splice sites can lead to partial inclusions or exclusions.

Targets of U11/U12 snRNPs can also be blocked. Splice modification can be conveniently assayed by reverse-transcriptase polymerase chain reaction (RT-PCR) and is seen as a band shift after gel electrophoresis of RT-PCR products.

Other Applications: Blocking Other mRNA Sites, Use as Probes, etc.

Morpholinos have been used to block miRNA activity and maturation. Fluorescein tagged Morpholinos, combined with fluorescein-specific antibodies, can be used as probes for in-situ hybridization to miRNAs. Morpholinos can block ribozyme activity. U2 and U12 snRNP functions have been inhibited by Morpholinos. Morpholinos targeted to "slippery" mRNA sequences within protein coding regions can induce translational frameshifts.

Morpholinos can block editing of RNA. Activities of Morpholinos against this variety of targets suggest that Morpholinos can be used as a general-purpose tool for blocking interactions of proteins or nucleic acids with mRNA.

Specificity, Stability and Non-Antisense Effects

Morpholinos have become a standard knockdown tool in animal embryonic systems, which have a broader range of gene expression than adult cells and can be strongly affected by an off-target interaction. Following initial injections into frog or fish embryos at the single-cell or few-cell stages, Morpholino effects can be measured up to five days later, after most of the processes of organogenesis and differentiation are past, with observed phenotypes consistent with target-gene knockdown. Control oligos with irrelevant sequences usually produce no change in embryonic phenotype, evidence of the Morpholino oligo's sequence-specificity and lack of non-antisense effects. The dose required for a knockdown can be reduced by coinjection of several Morpholino oligos targeting the same mRNA, which is an effective strategy for reducing or eliminating dose-dependent off-target RNA interactions.

mRNA rescue experiments can often restore the wild-type phenotype to the embryos and provide evidence for the specificity of a Morpholino. In an mRNA rescue, a Morpholino is co-injected with an mRNA that codes for the same protein that the Morphlino is targeted to knock down. However, the rescue mRNA has a modified 5'-UTR (untranslated region) so that the rescue mRNA contains no target for the Morpholino but the rescue mRNA's coding region encodes the protein of interest. Translation of the rescue mRNA replaces production of the protein that was knocked down by the Morpholino. Since the rescue mRNA would not affect phenotypic changes due to modulation of off-target gene expression by the Morpholino, this return to wild-type phenotype is further evidence of Morpholino specificity.

Because of their completely unnatural backbones, Morpholinos are not recognised by cellular proteins. Nucleases do not degrade Morpholinos, nor are they degraded in serum or in cells. Morpholinos do not activate toll-like receptors and so they do not activate innate immune responses such as interferon induction or the NF-êB-mediated inflammation response. Morpholinos are not known to modify methylation of DNA.

Up to 18% of Morpholinos appear to induce nontarget-related phenotypes including cell death in the central nervous system and somite tissues of zebrafish embryos. Most of these effects have been shown to be due to activation of p53-mediated apoptosis, and can be suppressed by co-injection of an anti-p53 Morpholino along with the experimental Morpholino; moreover, the p53-mediated apoptotic effect of a Morpholino knockdown has been phenocopied using another antisense structural type, showing the p53-mediated apoptosis to be a consequence of the loss of the targeted protein and not a consequence of the oligo type used to knock down the protein. It appears that these effects are sequence-specific; as in most cases, if a Morpholino is associated with non-target effects, the 4-base mismatch Morpholino will not trigger these effects.

A cause for concern in the use of Morpholinos is the potential for "off-target" effects. The question of whether an observed morphant phenotype is due to the intended knockdown or an interaction with an off-target RNA can often be addressed by running another experiment to confirm that the observed morphant phenotype results from the knockdown of the expected target. This can be done by recapitulating the morphant phenotype with a second, non-overlapping Morpholino targeting the same mRNA or by confirmation of the observed phenotypes by use of a mutant strain or dominant-negative methods. As mentioned above, rescue of observed phenotypes by coinjecting a rescue mRNA is, when feasible, a reliable test of specificity of a Morpholino.

Delivery

For a Morpholino to be effective, it must be delivered past the cell membrane into the cytosol of a cell. Once in the cytosol, Morpholinos freely diffuse between the cytosol and nucleus, as demonstrated by the nuclear splice-modifying activity of Morpholinos observed after microinjection into the cytosol of cells. Different methods are used for delivery into embryos, into cultured cells or into adult animals. A microinjection apparatus is usually used for delivery into an embryo, with injections most commonly performed at the single-cell or few-cell stage; an alternative method for embryonic delivery is electroporation, which can deliver oligos into tissues of later embryonic stages. Common techniques for delivery into cultured cells include the Endo-Porter peptide (which causes the Morpholino to be released from endosomes), the Special Delivery system (no longer commercially available, used a Morpholino-DNA heteroduplex and an ethoxylated polyethylenimine delivery reagent), electroporation, or scrape loading.

Delivery into adult tissues is usually difficult, though there are a few systems allowing useful uptake of unmodified Morpholino oligos (including the inherently leaky muscle cells caused by Duchenne muscular dystrophy or the vascular endothelial cells stressed during balloon angioplasty). Though they permeate through intercellular spaces in tissues effectively, unconjugated PMOs have limited distribution into the cytosol and nuclear spaces within healthy tissues following IV administration.

Systemic delivery into many cells in adult organisms can be accomplished by using covalent conjugates of Morpholino oligos with cell-penetrating peptides, and, while toxicity has been associated with moderate doses of the peptide conjugates, they have been used *in vivo* for effective oligo delivery at doses below those causing observed toxicity. An octa-guanidinium dendrimer attached to the end of a Morpholino can deliver the modified oligo (called a Vivo-Morpholino) from the blood to the cytosol. Delivery-enabled Morpholinos, such as peptide conjugates and Vivo-Morpholinos, show promise as therapeutics for viral and genetic diseases.

Peptide Nucleic Acid

Peptide nucleic acid (PNA) is an artificially synthesized polymer similar to DNA or RNA invented by Peter E. Nielsen (Univ. Copenhagen), Michael Egholm (Univ. Copenhagen), Rolf H. Berg (Risø National Lab), and Ole Buchardt (Univ. Copenhagen) in 1991. The name is somewhat of a misnomer as PNA is not an acid.

Synthetic peptide nucleic acid oligomers have been used in recent years in molecular biology procedures, diagnostic assays and antisense therapies. Due to their higher binding strength it is not necessary to design long PNA oligomers for use in these roles, which usually require oligonucleotide probes of 20–25 bases. The main concern of the length of the PNA-oligomers is to guarantee the specificity. PNA oligomers also show greater specificity in binding to complementary DNAs, with a PNA/DNA base mismatch being more destabilizing than a similar mismatch in a DNA/DNA duplex. This binding strength and specificity also applies to PNA/RNA duplexes. PNAs are not easily recognised by either nucleases or proteases, making them resistant to enzyme degradation. PNAs are also stable over a wide pH range. Though an unmodified PNA cannot readily cross cell membranes to enter the cytosol, covalently coupling a cell penetrating peptide to a PNA can improve cytosolic delivery.

PNA is not known to occur naturally but N-(2-aminoethyl)-glycine (AEG), the backbone of PNA, has been found produced by cyanobacteria.

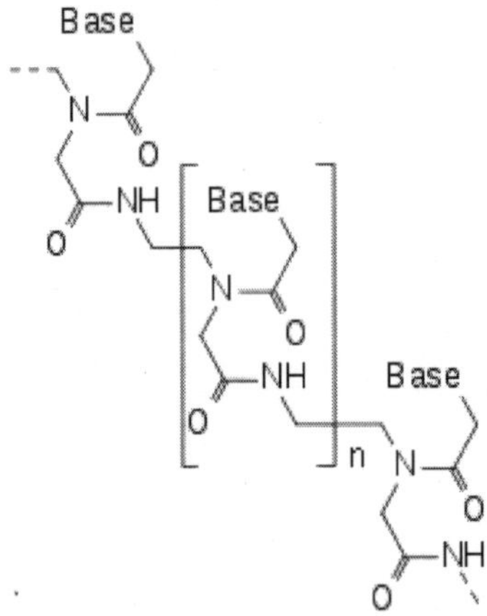

DNA and RNA have a deoxyribose and ribose sugar backbone, respectively, whereas PNA's backbone is composed of repeating N-(2-aminoethyl)-glycine units linked by peptide bonds. The various purine and pyrimidine bases are linked to the backbone by a methylene bridge (-CH2-) and a carbonyl group (-(C=O)-). PNAs are depicted like peptides, with the N-terminus at the first (left) position and the C-terminus at the last (right) position.

PNA Bindings

Since the backbone of PNA contains no charged phosphate groups, the binding between PNA/DNA strands is stronger than between DNA/DNA strands due to the lack of electrostatic repulsion. Unfortunately, this also causes it to be rather hydrophobic, which makes it difficult to deliver to body cells in solution without being flushed out of the body first. Early experiments with homopyrimidine strands (strands consisting of only one repeated pyrimidine base) have shown that the T_m ("melting" temperature) of a 6-base thymine PNA/adenine DNA double helix was 31 °C in comparison to an equivalent 6-base DNA/DNA duplex that denatures at a temperature less than 10 °C. Mixed base PNA molecules are true mimics of DNA molecules in terms of base-pair recognition. PNA/PNA binding is stronger than PNA/DNA binding.

PNA World Hypothesis

It has been hypothesized that the earliest life on Earth may have used PNA as a genetic material due to its extreme robustness, simpler formation and possible spontaneous polymerization at 100°C (while water at standard pressure boils at this temperature, water at high pressure—as in deep ocean—boils at higher temperatures). If this is so, life evolved to a DNA/RNA-based system only at a later stage. Evidence for this PNA world hypothesis is however far from conclusive.

Applications

- Alteration of gene expression - both as inhibitor and promoter in different cases
- Antigene and antisense therapeutic agent
- Anticancer agent
- Antiviral, Antibacterial and Antiparasitic agent
- Molecular tools and probes of biosensor
- Detection of DNA sequences

Locked Nucleic Acid

A locked nucleic acid (LNA), often referred to as inaccessible RNA, is a modified RNA nucleotide. The ribose moiety of an LNA nucleotide is modified with an extra bridge connecting the 2' oxygen and 4' carbon. The bridge "locks" the ribose in the 3'-*endo* (North) conformation, which is often found in the A-form duplexes. LNA nucleotides can be mixed with DNA or RNA residues in the oligonucleotide whenever desired. Such oligomers are synthesized chemically and are commercially available. The locked ribose conformation enhances base stacking and backbone pre-organisation. This significantly increases the hybridization properties (melting temperature) of oligonucleotides. LNA was independently synthesized by the group of Jesper Wengel in 1998, soon after the first synthesis by the group of Takeshi Imanishi in 1997. The exclusive rights to the LNA technology were secured in 1997 by Exiqon A/S, a Danish biotech company.

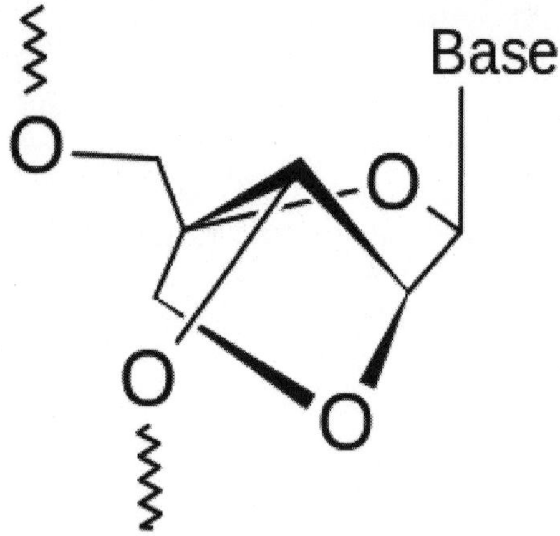

Figure: *Conformational structure of an LNA monomer â-D configuration*

LNA nucleotides are used to increase the sensitivity and specificity of expression in DNA microarrays, FISH probes, quantitative PCR probes and other molecular biology techniques based on oligonucleotides. For the *in situ* detection of miRNA the use of LNA is currently (2005) the only efficient method. A triplet of LNA nucleotides surrounding a single-base mismatch site maximizes LNA probe specificity unless the probe contains the guanine base of G-T mismatch.

Using LNA based oligonucleotides therapeutically is an emerging field in biotechnology. The Danish pharmaceutical company Santaris Pharma a/s owns the sole rights to therapeutic uses of LNA technology, and is now developing a new, LNA based, hepatitis C drug called miravirsen, targeting miR-122, which is in Phase II clinical testing as of late 2010.

An LNA-specific molecular dynamics (MD) force field has been created which is able to accurately model LNA MD.

Benefits of the LNA Technology

Some of the benefits of using LNA include:

- Ideal for the detection of short RNA and DNA targets
- Increases the thermal stability of duplexes
- Capable of single nucleotide discrimination
- Resistant to exo- and endonucleases resulting in high stability in vivo and in vitro applications

- Increased target specificity
- Facilitates Tm normalization
- Strand invasion properties enables detection of "hard to access" samples
- Compatible with standard enzymatic processes

Applications of the LNA Technology

Some proven applications of LNA include:

- Allele-specific PCR: allows for the design of shorter primers, without compromising binding specificity
- Microarray gene expression profiling: provides increased sensitivity and selectivity with smaller amounts of substrates
- Small RNA research
- SNP genotyping
- mRNA antisense oligonucleotides
- RNAi
- DNAzymes
- Fluorescence Polarization probes
- Molecular Beacons
- Gene repair/exon skipping
- Splice variant detection
- Comparative genome hybridization (GCH)

Other therapeutic and diagnostic applications of LNA technology are in development. A cooperation between Southern University of Denmark and Decan University are focusing on using LNA to diagnose cancer-stamcells, which will be a major leap forward to avoid cancer-patients having relapse. The team filed patent in february 2014 and will continue with trials on rodents in late 2014.

4

RNA Interference (Which Uses *Double-Strand* RNA)

RNA interference (RNAi) is a biological process in which RNA molecules inhibit gene expression, typically by causing the destruction of specific mRNA molecules. Historically, it was known by other names, including *co-suppression*, *post transcriptional gene silencing* (PTGS), and *quelling*. Only after these apparently unrelated processes were fully understood did it become clear that they all described the RNAi phenomenon. Andrew Fire and Craig C. Mello shared the 2006 Nobel Prize in Physiology or Medicine for their work on RNA interference in the nematode worm *Caenorhabditis elegans*, which they published in 1998.

Two types of small ribonucleic acid (RNA) molecules – microRNA (miRNA) and small interfering RNA (siRNA) – are central to RNA interference. RNAs are the direct products of genes, and these small RNAs can bind to other specific messenger RNA (mRNA) molecules and either increase or decrease their activity, for example by preventing an mRNA from producing a protein. RNA interference has an important role in defending cells against parasitic nucleotide sequences – viruses and transposons. It also influences development.

The RNAi pathway is found in many eukaryotes including animals and is initiated by the enzyme Dicer, which cleaves long double-stranded RNA (dsRNA) molecules into short double stranded fragments of ~20 nucleotide siRNAs. Each siRNA is unwound into two single-stranded (ss) ssRNAs, respectively the passenger strand and the guide strand. The passenger strand is degraded and the guide strand is incorporated into the RNA-induced silencing complex (RISC). The most well-studied outcome is post-transcriptional gene silencing, which occurs when the guide strand pairs with a complementary sequence in a messenger RNA molecule and induces cleavage by Argonaute, the catalytic component of the RISC complex. In some organisms, this process spreads systemically, despite the initially limited molar concentrations of siRNA.

RNAi is a valuable research tool, both in cell culture and in living organisms, because synthetic dsRNA introduced into cells can selectively and robustly induce suppression of specific genes of interest. RNAi may be used for large-scale screens that systematically shut down each gene in the cell, which can help to identify the components necessary for a particular cellular process or an event such as cell division. The pathway is also used as a practical tool in biotechnology, medicine and insecticides.

Cellular Mechanism

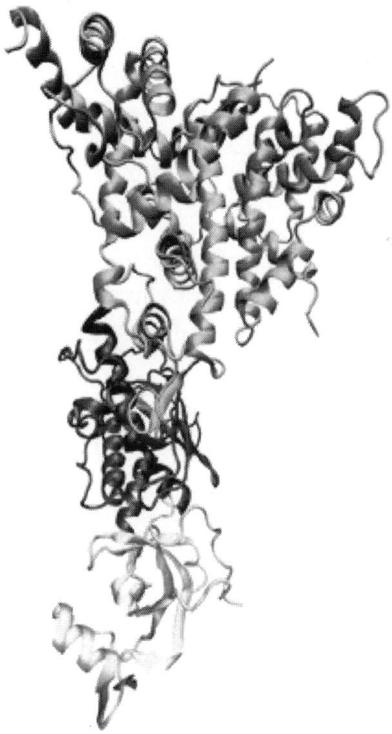

Figure: *The dicer protein from* Giardia intestinalis, *which catalyzes the cleavage of dsRNA to siRNAs. The RNase domains are coloured green, the PAZ domain yellow, the platform domain red, and the connector helix blue.*

RNAi is an RNA-dependent gene silencing process that is controlled by the RNA-induced silencing complex (RISC) and is initiated by short double-stranded RNA molecules in a cell's cytoplasm, where they interact with the catalytic RISC component argonaute. When the dsRNA is exogenous (coming from infection by a virus with an RNA genome or laboratory manipulations), the RNA is imported directly into the cytoplasm and cleaved to short fragments by Dicer. The initiating dsRNA can also be endogenous (originating in the cell), as in pre-microRNAs expressed from RNA-coding genes in the genome. The primary transcripts from such genes are first processed to form the characteristic stem-loop structure of pre-miRNA in the nucleus, then exported to the cytoplasm. Thus, the two dsRNA pathways, exogenous and endogenous, converge at the RISC complex.

dsRNA Cleavage

Endogenous dsRNA initiates RNAi by activating the ribonuclease protein Dicer, which binds and cleaves double-stranded RNAs (dsRNAs) to produce double-stranded fragments of 20–25 base pairs with a 2-nucleotide overhang at the 3' end. Bioinformatics studies on the genomes of multiple organisms suggest this length maximizes target-gene specificity and minimizes non-specific effects. These short double-stranded fragments are called small interfering RNAs (siRNAs). These siRNAs are then separated into single strands and integrated into an active RISC complex. After integration into the RISC, siRNAs base-pair to their target mRNA and cleave it, thereby preventing it from being used as a translation template.

Exogenous dsRNA is detected and bound by an effector protein, known as RDE-4 in *C. elegans* and R2D2 in *Drosophila*, that stimulates dicer activity. This protein only binds long dsRNAs, but the mechanism producing this length specificity is unknown. This RNA-binding protein then facilitates the transfer of cleaved siRNAs to the RISC complex.

In *C. elegans* this initiation response is amplified through the synthesis of a population of 'secondary' siRNAs during which the dicer-produced initiating or 'primary' siRNAs are used as templates. These 'secondary' siRNAs are structurally distinct from dicer-produced siRNAs and appear to be produced by an RNA-dependent RNA polymerase (RdRP).

MicroRNA

MicroRNAs (miRNAs) are genomically encoded non-coding RNAs that help regulate gene expression, particularly during development. The phenomenon of RNA interference, broadly defined, includes the endogenously induced gene silencing effects of miRNAs as well as silencing triggered by foreign dsRNA. Mature miRNAs are structurally similar to siRNAs produced from exogenous dsRNA, but before reaching maturity, miRNAs must first undergo extensive post-transcriptional modification. An miRNA is expressed from a much longer RNA-coding gene as a primary transcript known as a *pri-miRNA* which is processed, in the cell nucleus, to a 70-nucleotide stem-loop structure called a *pre-miRNA* by the microprocessor complex. This complex consists of an RNase III enzyme called Drosha and a dsRNA-binding protein DGCR8. The dsRNA portion of this pre-miRNA is bound and cleaved by Dicer to produce the mature miRNA molecule that can be integrated into the RISC complex; thus, miRNA and siRNA share the same downstream cellular machinery downstream.

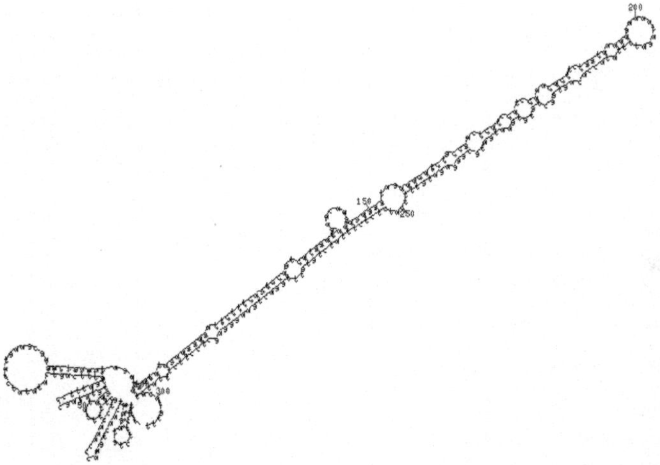

Figure: *The stem-loop secondary structure of a pre-microRNA from* Brassica oleracea.

siRNAs derived from long dsRNA precursors differ from miRNAs in that miRNAs, especially those in animals, typically have incomplete base pairing to a target and inhibit the translation of many different mRNAs with similar sequences. In contrast, siRNAs typically base-pair perfectly and induce mRNA cleavage only in a single, specific target. In *Drosophila* and *C. elegans*, miRNA and siRNA are processed by distinct argonaute proteins and dicer enzymes.

RISC Activation and Catalysis

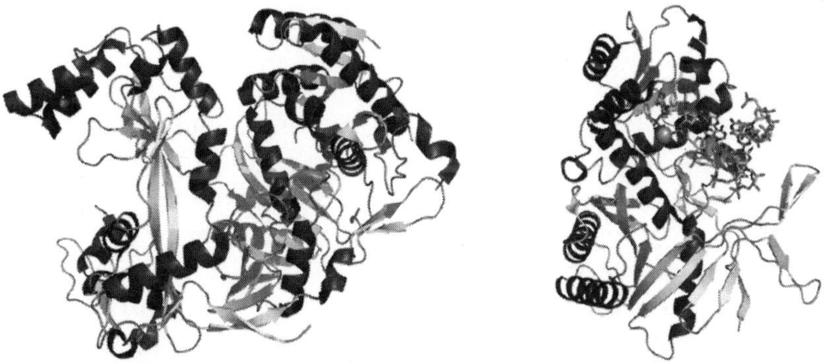

Figure: Left: *A full-length argonaute protein from the archaea species* Pyrococcus furiosus. Right: *The PIWI domain of an argonaute protein in complex with double-stranded RNA.*

The active components of an RNA-induced silencing complex (RISC) are endonucleases called argonaute proteins, which cleave the target mRNA strand complementary to their bound siRNA. As the fragments produced by dicer are double-stranded, they could each in theory produce a functional siRNA. However, only one of the two strands, which is known as the *guide strand*, binds the argonaute protein and directs gene silencing.

The other *anti-guide strand* or *passenger strand* is degraded during RISC activation. Although it was first believed that an ATP-dependent helicase separated these two strands, the process proved to be ATP-independent and performed directly by the protein components of RISC. However, an *in vitro* kinetic analysis of RNAi in the presence and absence of ATP showed that ATP may be required to unwind and remove the cleaved mRNA strand from the RISC complex after catalysis. The guide strand tends to be the one whose 5' end is less stably paired to its complement, but strand selection is unaffected by the direction in which dicer cleaves the dsRNA before RISC incorporation. Instead, the R2D2 protein may serve as the differentiating factor by binding the more-stable 5' end of the passenger strand.

The structural basis for binding of RNA to the argonaute protein was examined by X-ray crystallography of the binding domain of an RNA-bound argonaute protein. Here, the phosphorylated 5' end of the RNA strand enters a conserved basic surface pocket and makes contacts through a divalent cation (an atom with two positive charges) such as magnesium and by aromatic stacking (a process that allows more than one atom to share an electron by passing it back and forth) between the 5' nucleotide in the siRNA and a conserved tyrosine residue.

This site is thought to form a nucleation site for the binding of the siRNA to its mRNA target. Analysis of the inhibitory effect of mismatches in either the 5' or 3' end of the guide strand has demonstrated that the 5' end of the guide strand is likely responsible for matching and binding the target mRNA, while the 3' end is responsible for physically arranging target mRNA into a cleavage-favourable RISC region.

It is not understood how the activated RISC complex locates complementary mRNAs within the cell. Although the cleavage process has been proposed to be linked to translation, translation of the mRNA target is not essential for RNAi-mediated degradation. Indeed, RNAi may be more effective against mRNA targets that are not translated. Argonaute proteins are localized to specific regions in the cytoplasm called P-bodies (also cytoplasmic bodies or GW bodies), which are regions with high rates of mRNA decay; miRNA activity is also clustered in P-bodies. Disruption of P-bodies decreases the efficiency of RNA interference, suggesting that they are a critical site in the RNAi process.

Transcriptional Silencing

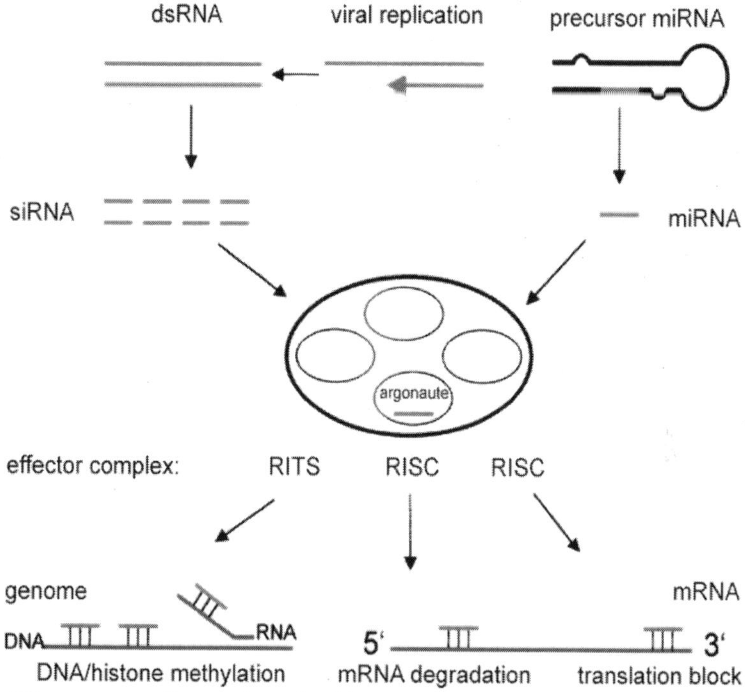

Figure: *The enzyme dicer trims double stranded RNA, to form small interfering RNA or microRNA. These processed RNAs are incorporated into the RNA-induced silencing complex (RISC), which targets messenger RNA to prevent translation.*

Components of the RNAi pathway are used in many eukaryotes in the maintenance of the organisation and structure of their genomes. Modification of histones and associated induction of heterochromatin formation serves to downregulate genes pre-transcriptionally; this process is referred to as RNA-induced transcriptional silencing (RITS), and is carried out by a complex of proteins called the RITS complex. In fission yeast this complex contains argonaute, a chromodomain protein Chp1, and a protein called Tas3 of unknown function. As a consequence, the induction and spread of heterochromatic regions requires the argonaute and RdRP proteins. Indeed, deletion of these genes in the fission yeast *S. pombe* disrupts histone methylation and centromere formation, causing slow or stalled anaphase during cell division. In some cases, similar processes associated with histone modification have been observed to transcriptionally upregulate genes.

The mechanism by which the RITS complex induces heterochromatin formation and organisation is not well understood. Most studies have focused on the mating-type region in fission yeast, which may not be representative of activities in other genomic regions/organisms. In maintenance of existing heterochromatin regions, RITS forms a complex with siRNAs complementary to the local genes and stably binds local methylated histones, acting co-transcriptionally to degrade any nascent pre-mRNA transcripts that are initiated by RNA polymerase. The formation of such a heterochromatin region, though not its maintenance, is dicer-dependent, presumably because dicer is required to generate the initial complement of siRNAs that target subsequent transcripts. Heterochromatin maintenance has been suggested to function as a self-reinforcing feedback loop, as new siRNAs are formed from the occasional nascent transcripts by RdRP for incorporation into local RITS complexes. The relevance of observations from fission yeast mating-type regions and centromeres to mammals is not clear, as heterochromatin maintenance in mammalian cells may be independent of the components of the RNAi pathway.

Crosstalk with RNA Editing

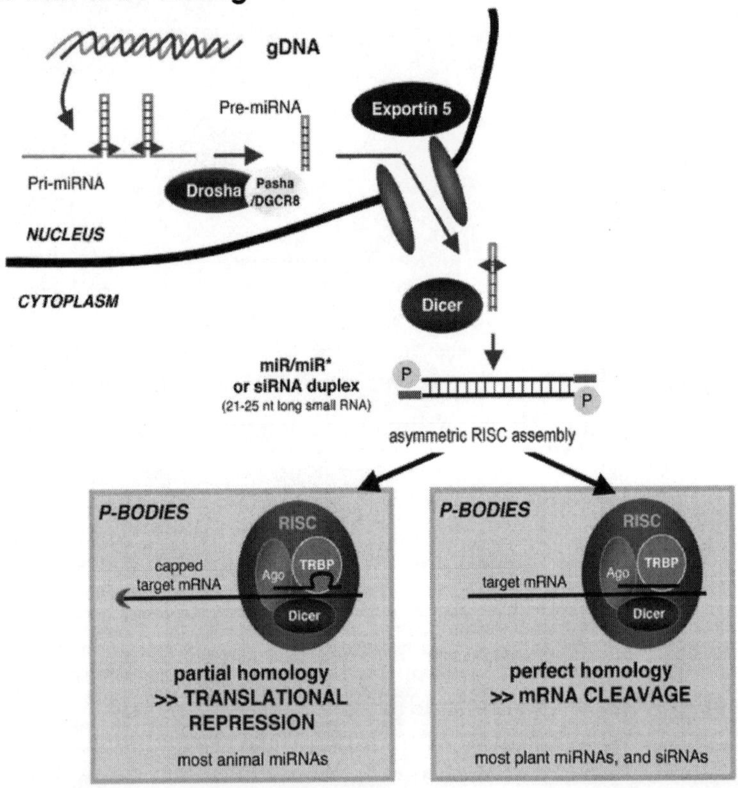

Figure: Illustration of the major differences between plant and animal gene silencing. Natively expressed microRNA or exogenous small interfering RNA is processed by dicer and integrated into the RISC complex, which mediates gene silencing.

The type of RNA editing that is most prevalent in higher eukaryotes converts adenosine nucleotides into inosine in dsRNAs via the enzyme adenosine deaminase (ADAR). It was originally proposed in 2000 that the RNAi and A'!I RNA editing pathways might compete for a common dsRNA substrate. Some pre-miRNAs do undergo A'!I RNA editing and this mechanism may regulate the processing and expression of mature miRNAs. Furthermore, at least one mammalian ADAR can sequester siRNAs from RNAi pathway components. Further support for this model comes from studies on ADAR-null *C. elegans* strains indicating that A'!I RNA editing may counteract RNAi silencing of endogenous genes and transgenes.

Variation among Organisms

Organisms vary in their ability to take up foreign dsRNA and use it in the RNAi pathway. The effects of RNA interference can be both systemic and heritable in plants and *C. elegans*, although not in *Drosophila* or mammals. In plants, RNAi is thought to propagate by the transfer of siRNAs between cells through plasmodesmata (channels in the cell walls that enable communication and transport). Heritability comes from methylation of promoters targeted by RNAi; the new methylation pattern is copied in each new generation of the cell. A broad general distinction between plants and animals lies in the targeting of endogenously produced miRNAs; in plants, miRNAs are usually perfectly or nearly perfectly complementary to their target genes and induce direct mRNA cleavage by RISC, while animals' miRNAs tend to be more divergent in sequence and induce translational repression. This translational effect may be produced by inhibiting the interactions of translation initiation factors with the messenger RNA's polyadenine tail.

Some eukaryotic protozoa such as *Leishmania major* and *Trypanosoma cruzi* lack the RNAi pathway entirely. Most or all of the components are also missing in some fungi, most notably the model organism *Saccharomyces cerevisiae*. The presence of RNAi in other budding yeast species such as *Saccharomyces castellii* and *Candida albicans*, further demonstrates that inducing two RNAi-related proteins from *S. castellii* facilitates RNAi in *S. cerevisiae*. That certain ascomycetes and basidiomycetes are missing RNA interference pathways indicates that proteins required for RNA silencing have been lost independently from many fungal lineages, possibly due to the evolution of a novel pathway with similar function, or to the lack of selective advantage in certain niches.

Related Prokaryotic Systems

Gene expression in prokaryotes is influenced by an RNA-based system similar in some respects to RNAi. Here, RNA-encoding genes control mRNA abundance or translation by producing a complementary RNA that anneals to an mRNA. However these regulatory RNAs are not generally considered to be analogous to miRNAs because the dicer enzyme is not involved. It has been suggested that CRISPR interference systems in prokaryotes are analogous to eukaryotic RNA interference systems, although none of the protein components are orthologous.

Biological Functions

Immunity

RNA interference is a vital part of the immune response to viruses and other foreign genetic material, especially in plants where it may also prevent the self-propagation of transposons. Plants such as *Arabidopsis thaliana* express multiple dicer homologs that are specialised to react differently when the plant is exposed to different viruses. Even before the RNAi pathway was fully understood, it was known that induced gene silencing in plants could spread throughout the plant in a systemic effect and could be transferred from stock to scion plants via grafting. This phenomenon has since been recognised as a feature of the plant adaptive immune system and allows the entire plant to respond to a virus after an initial localized encounter. In response, many plant viruses have evolved elaborate mechanisms to suppress the RNAi response. These include viral proteins that bind short double-stranded RNA fragments with single-stranded overhang ends, such as those produced by dicer. Some plant genomes also express endogenous siRNAs in response to infection by specific types of bacteria. These effects may be part of a generalized response to pathogens that downregulates any metabolic process in the host that aids the infection process.

Although animals generally express fewer variants of the dicer enzyme than plants, RNAi in some animals produces an antiviral response. In both juvenile and adult *Drosophila*, RNA interference is important in antiviral innate immunity and is active against pathogens such as Drosophila X virus. A similar role in immunity may operate in *C. elegans*, as argonaute proteins are upregulated in response to viruses and worms that overexpress components of the RNAi pathway are resistant to viral infection.

The role of RNA interference in mammalian innate immunity is poorly understood, and relatively little data is available. However, the existence of viruses that encode genes able to suppress the RNAi response in mammalian cells may be evidence in favour of an RNAi-dependent mammalian immune response, although this hypothesis has been challenged as poorly substantiated Maillard et al. and Li et al. provide evidence for the existence of a functional antiviral RNAi pathway in mammalian cells. Other functions for RNAi in mammalian viruses also exist, such as miRNAs expressed by the herpes virus that may act as heterochromatin organisation triggers to mediate viral latency

Downregulation of Genes

Endogenously expressed miRNAs, including both intronic and intergenic miRNAs, are most important in translational repression and in the regulation of development, especially on the timing of morphogenesis and the maintenance of undifferentiated or incompletely differentiated cell types such as stem cells

The role of endogenously expressed miRNA in downregulating gene expression was first described in *C. elegans* in 1993. In plants this function was discovered when the "JAW microRNA" of *Arabidopsis* was shown to be involved in the regulation of several genes that control plant shape. In plants, the majority of genes regulated by miRNAs are transcription factors; thus miRNA activity is particularly wide-ranging and regulates entire gene networks during development by modulating the expression of key regulatory genes, including transcription factors as well as F-box proteins. In many organisms, including humans, miRNAs are linked to the formation of tumors and dysregulation of the cell cycle. Here, miRNAs can function as both oncogenes and tumor suppressors.

Upregulation of Genes

RNA sequences (siRNA and miRNA) that are complementary to parts of a promoter can increase gene transcription, a phenomenon dubbed RNA activation. Part of the mechanism for how these RNA upregulate genes is known: dicer and argonaute are involved, possibly via histone demethylation. miRNAs have been proposed to upregulate their target genes upon cell cycle arrest, via unknown mechanisms.

Evolution

Based on parsimony-based phylogenetic analysis, the most recent common ancestor of all eukaryotes most likely already possessed an early RNA interference pathway; the absence of the pathway in certain eukaryotes is thought to be a derived characteristic. This ancestral RNAi system probably contained at least one dicer-like protein, one argonaute, one PIWI protein, and an RNA-dependent RNA polymerase that may also have played other cellular roles. A large-scale comparative genomics study likewise indicates that the eukaryotic crown group already possessed these components, which may then have had closer functional associations with generalized RNA degradation systems such as the exosome. This study also suggests that the RNA-binding argonaute protein family, which is shared among eukaryotes, most archaea, and at least some bacteria (such as *Aquifex aeolicus*), is homologous to and originally evolved from components of the translation initiation system.

The ancestral function of the RNAi system is generally agreed to have been immune defence against exogenous genetic elements such as transposons and viral genomes. Related functions such as histone modification may have already been present in the ancestor of modern eukaryotes, although other functions such as regulation of development by miRNA are thought to have evolved later.

RNA interference genes, as components of the antiviral innate immune system in many eukaryotes, are involved in an evolutionary arms race with viral genes. Some viruses have evolved mechanisms for suppressing the RNAi response in their host cells, particularly for plant viruses. Studies of evolutionary rates in *Drosophila* have shown that genes in the RNAi pathway are subject to strong directional selection and are among the fastest-evolving genes in the *Drosophila* genome.

Applications

Gene Knockdown

The RNA interference pathway is often exploited in experimental biology to study the function of genes in cell culture and *in vivo* in model organisms. Double-stranded RNA is synthesized with a sequence complementary to a gene of interest and introduced into a cell or organism, where it is recognised as exogenous genetic material and activates the RNAi pathway. Using this mechanism, researchers can cause a drastic decrease in the expression of a targeted gene. Studying the effects of this decrease can show the physiological role of the gene product. Since RNAi may not totally abolish expression of the gene, this technique is sometimes referred as a "knockdown", to distinguish it from "knockout" procedures in which expression of a gene is entirely eliminated.

Extensive efforts in computational biology have been directed toward the design of successful dsRNA reagents that maximize gene knockdown but minimize "off-target" effects. Off-target effects arise when an introduced RNA has a base sequence that can pair with and thus reduce the expression of multiple genes. Such problems occur more frequently when the dsRNA contains repetitive sequences. It has been estimated from studying the genomes of humans, *C. elegans* and *S. pombe* that about 10% of possible siRNAs have substantial off-target effects. A multitude of software tools have been developed implementing algorithms for the design of general mammal-specific, and virus-specific siRNAs that are automatically checked for possible cross-reactivity.

Depending on the organism and experimental system, the exogenous RNA may be a long strand designed to be cleaved by dicer, or short RNAs designed to serve as siRNA substrates. In most mammalian cells, shorter RNAs are used because long double-stranded RNA molecules induce the mammalian interferon response, a form of innate immunity that reacts nonspecifically to foreign genetic material. Mouse oocytes and cells from early mouse embryos lack this reaction to exogenous dsRNA and are therefore a common model system for studying mammalian gene-knockdown effects. Specialised laboratory techniques have also been developed to improve the utility of RNAi in mammalian systems by avoiding the direct introduction of siRNA, for example, by stable transfection with a plasmid encoding the appropriate sequence from which siRNAs can be transcribed, or by more elaborate lentiviral vector systems allowing the inducible activation or deactivation of transcription, known as *conditional RNAi.*

Functional Genomics

Most functional genomics applications of RNAi in animals have used *C. elegans* and *Drosophila*, as these are the common model organisms in which RNAi is most effective. *C. elegans* is particularly useful for RNAi research for two reasons: firstly, the effects of gene silencing are generally heritable, and secondly because delivery of the dsRNA is extremely simple. Through a mechanism whose details are poorly understood, bacteria such as *E. coli* that carry the desired dsRNA can be fed to the worms and will transfer their RNA payload to the worm via the intestinal tract. This "delivery by feeding" is just as effective at inducing gene silencing as more costly and time-consuming delivery methods, such as soaking the worms in dsRNA solution and injecting dsRNA into the gonads. Although delivery is more difficult in most other organisms, efforts are also underway to undertake large-scale genomic screening applications in cell culture with mammalian cells. Approaches to the design of genome-wide RNAi libraries can require more sophistication than the design of a single siRNA for a defined set of experimental conditions. Artificial neural networks are frequently used to design siRNA libraries and to predict their likely efficiency at gene knockdown. Mass genomic screening is widely seen as a promising method for genome annotation and has triggered the development of high-throughput screening methods based on microarrays. However, the utility of these screens and the ability of techniques developed on model organisms to generalize to even closely related species has been questioned, for example from *C. elegans* to related parasitic nematodes.

Functional genomics using RNAi is a particularly attractive technique for genomic mapping and annotation in plants because many plants are polyploid, which presents substantial challenges for more traditional genetic engineering methods. For example, RNAi has been successfully used for functional genomics studies in bread wheat (which is hexaploid) as well as more common plant model systems *Arabidopsis* and maize.

Medicine

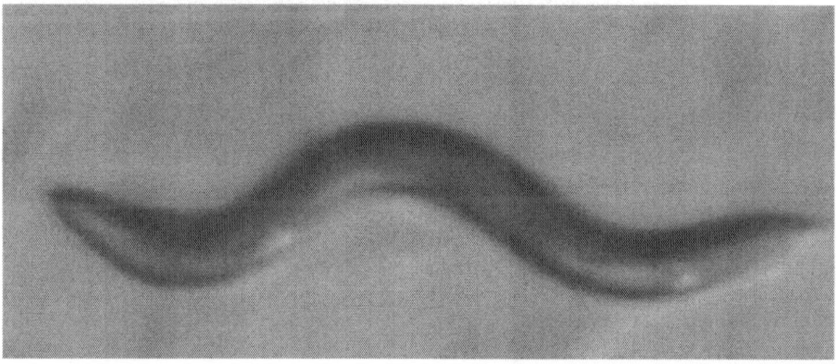

Figure: *An adult* C. elegans *worm, grown under RNAi suppression of a nuclear hormone receptor involved in desaturase regulation. These worms have abnormal fatty acid metabolism but are viable and fertile.*

It may be possible to exploit RNA interference in therapy. Although it is difficult to introduce long dsRNA strands into mammalian cells due to the interferon response, the use of short interfering RNA has been more successful. Among the first applications to reach clinical trials were in the treatment of macular degeneration and respiratory syncytial virus. RNAi has also been shown to be effective in reversing induced liver failure in mouse models.

Antiviral

Potential antiviral therapies include topical microbicide treatments that use RNAi to treat infection (at Harvard Medical School; in mice, so far) by herpes simplex virus type 2 and the inhibition of viral gene expression in cancerous cells, knockdown of host receptors and coreceptors for HIV, the silencing of hepatitis A and hepatitis B genes, silencing of influenza gene expression, and inhibition of measles viral replication. Potential treatments for neurodegenerative diseases have also been proposed, with particular attention to polyglutamine diseases such as Huntington's disease.

RNA interference-based applications are being developed to target persistent HIV-1 infection. Viruses like HIV-1 are particularly difficult targets for RNAi-attack because they are escape-prone, which requires combinatorial RNAi strategies to prevent viral escape.

Cancer

RNA interference is also a promising way to treat cancers by silencing genes differentially upregulated in tumor cells or genes involved in cell division. A key area of research in the use of RNAi for clinical applications is the development of a safe delivery method, which to date has involved mainly viral vector systems similar to those suggested for gene therapy.

Due to safety concerns with viral vectors, nonviral delivery methods, typically employing lipid-based or polymeric vectors, are also promising candidates. Computational modelling of nonviral siRNA delivery paired with 'in vitro *and* in vivo *gene knockdown studies elucidated the temporal behaviour of RNAi in these systems. The model used an input bolus dose of siRNA and computationally and experimentally showed that knockdown duration was dependent mainly on the doubling time of the cells to which siRNA was delivered, while peak knockdown depended primarily on the delivered dose. Kinetic considerations of RNAi are imperative to safe and effective dosing schedules as nonviral methods of inducing RNAi continue to be developed.*

Safety

Despite the proliferation of promising cell culture studies for RNAi-based drugs, some concern has been raised regarding the safety of RNA interference, especially the potential for "off-target" effects in which a gene with a coincidentally similar sequence to the targeted gene is also repressed. A computational genomics study estimated that the error rate of off-target interactions is about 10%.

One major study of liver disease in mice reported that 23 out of 49 distinct RNAi treatment protocols resulted in death. Researchers hypothesized this alarmingly high rate to be the result of "oversaturation" of the dsRNA pathway, due to the use of shRNAs that have to be processed in the nucleus and exported to the cytoplasm using an active mechanism. Such considerations are under active investigation, to reduce their impact in the potential therapeutic applications.

RNAi in vivo delivery to tissues still eludes science—especially to tissues deep within the body. RNAi delivery is only easily accessible to surface tissues such as the eye and respiratory tract. In these instances, siRNA has been used in direct contact with the tissue for transport. The resulting RNAi successfully focused on target genes. When delivering siRNA to deep tissues, the siRNA must be protected from nucleases, but targeting specific areas becomes the main difficulty. This difficulty has been combatted with high dosage levels of siRNA to ensure the tissues have been reached, however in these cases hepatotoxicity was reported.

Biotechnology

RNA interference has been used for applications in biotechnology and is nearing commercialization in others.

Foods

RNAi has been used to genetically engineer plants to produce lower levels of natural plant toxins. Such techniques take advantage of the stable and heritable RNAi phenotype in plant stocks. Cotton seeds are rich in dietary protein but naturally contain the toxic terpenoid product gossypol, making them unsuitable for human consumption. RNAi has been used to produce cotton stocks whose seeds contain reduced levels of delta-cadinene synthase, a key enzyme in gossypol production, without affecting the enzyme's production in other parts of the plant, where gossypol is itself important in preventing damage from plant pests. Similar efforts have been directed toward the reduction of the cyanogenic natural product linamarin in cassava plants.

No plant products that use RNAi-based genetic engineering have yet exited the experimental stage. Development efforts have successfully reduced the levels of allergens in tomato plants and fortification of plants such as tomatoes with dietary antioxidants. Previous commercial products, including the Flavr Savr tomato and two cultivars of ringspot-resistant papaya, were originally developed using antisense technology but likely exploited the RNAi pathway.

Other Crops

Another effort decreased the precursors of likely carcinogens in tobacco plants. Other plant traits that have been engineered in the laboratory include the production of non-narcotic natural products by the opium poppy and resistance to common plant viruses.

Insecticide

RNAi is under development as an insecticide, employing multiple approaches, including genetic engineering and topical application. Cells in the midgut of many larvae take up the molecules and help spread the signal throughout the insect's body.

RNAi has varying effects in different species of Lepidoptera (butterflies and moths). Possibly because their saliva is better at breaking down RNA, the cotton bollworm, the beet armyworm and the Asiatic rice borer have so far not been proven susceptible to RNAi by feeding.

To develop resistance to RNAi, the western corn rootworm would have to change the genetic sequence of its Snf7 gene at multiple sites. Combining multiple strategies, such as engineering the protein Cry, derived from a bacterium called Bacillus thuringiensis (Bt), and RNAi in one plant delay the onset of resistance.

One unconfirmed 2012 paper detected small RNAs from food plants in the blood of mice and humans. The consequences of RNA insecticides in the human bloodstream have not been investigated. Biological barriers—including saliva and blood enzymes and stomach acids may break down any ingested RNA. Critics charge that the human equivalent of the mouse diet in the study would be 33 kilograms of cooked rice a day. Two 2013 studies failed to detect RNAs in humans. Athletes consuming a diet of apples and bananas and monkeys consuming a fruit shake both appeared to be RNA-free.

Transgenic Plants

Transgenic crops have been made to express small bits of RNA, carefully chosen to silence crucial genes in target pests. RNAs exist that affect only insects that have specific genetic sequences. In 2009 a study showed RNAs that could kill any one of four fruit fly species while not harming the other three.

In 2012 Syngenta bought Belgian RNAi firm Devgen for $522 million and Monsanto paid $29.2 million for the exclusive rights to intellectual property from Alnylam Pharmaceuticals. The International Potato Centre in Lima, Peru is looking for genes to target in the sweet potato weevil, a beetle whose larvae ravage sweet potatos globally. Other researchers are trying to silence genes in ants, caterpillars and pollen beetles. Monsanto will likely be first to market, with a transgenic corn seed that expresses dsRNA based on gene Snf7 from the western corn rootworm, a beetle whose larvae annually cause one billion dollars in damage in the United States alone. A 2012 paper showed that silencing Snf7 stunts larval growth, killing them within days. In 2013 the same team showed that the RNA affects very few other species.

Topical

Alternatively dsRNA can be supplied without genetic engineering. One approach is to add them to irrigation water. The molecules are absorbed into the plants' vascular system and poison insects feeding on them. Another approach involves spraying RNA like a conventional pesticide. This would allow faster adaptation to resistance. Such approaches would require low cost sources of RNAs that do not currently exist.

Genome-Scale Screening

Genome-scale RNAi research relies on high-throughput screening (HTS) technology. RNAi HTS technology allows genome-wide loss-of-function screening and is broadly used in the identification of genes associated with specific phenotypes. This technology has been hailed as the second genomics wave, following the first genomics wave of gene expression microarray and single nucleotide polymorphism discovery platforms. One major advantage of genome-scale RNAi screening is its ability to simultaneously interrogate thousands of genes. With the ability generate a large amount of data per experiment, genome-scale RNAi screening has led to an explosion data generation rates. Exploiting such large data sets is a fundamental challenge, requiring suitable statistics/bioinformatics methods. The basic process of cell-based RNAi screening includes the choice of an RNAi library, robust and stable cell types, transfection with RNAi agents, treatment/incubation, signal detection, analysis and identification of important genes or therapeutical targets.

History

Figure: *Example petunia plants in which genes for pigmentation are silenced by RNAi. The left plant is wild-type; the right plants contain transgenes that induce suppression of both transgene and endogenous gene expression, giving rise to the unpigmented white areas of the flower.*

The discovery of RNAi was preceded first by observations of transcriptional inhibition by antisense RNA expressed in transgenic plants, and more directly by reports of unexpected outcomes in experiments performed by plant scientists in the United States and the Netherlands in the early 1990s. In an attempt to alter flower colours in petunias, researchers introduced additional copies of a gene encoding chalcone synthase, a key enzyme for flower pigmentation into petunia plants of normally pink or violet flower colour. The overexpressed gene was expected to result in darker flowers, but instead produced less pigmented, fully or partially white flowers, indicating that the activity of chalcone synthase had been substantially decreased; in fact, both the endogenous genes and the transgenes were downregulated in the white flowers. Soon after, a related event termed *quelling* was noted in the fungus *Neurospora crassa*, although it was not immediately recognised as related. Further investigation of the phenomenon in plants indicated that the downregulation was due to post-transcriptional inhibition of gene expression via an increased rate of mRNA degradation. This phenomenon was called *co-suppression of gene expression*, but the molecular mechanism remained unknown.

Not long after, plant virologists working on improving plant resistance to viral diseases observed a similar unexpected phenomenon. While it was known that plants expressing virus-specific proteins showed enhanced tolerance or resistance to viral infection, it was not expected that plants carrying only short, non-coding regions of viral RNA sequences would show similar levels of protection. Researchers believed that viral RNA produced by transgenes could also inhibit viral replication. The reverse experiment, in which short sequences of plant genes were introduced into viruses, showed that the targeted gene was suppressed in an infected plant. This phenomenon was labelled "virus-induced gene silencing" (VIGS), and the set of such phenomena were collectively called *post transcriptional gene silencing*.

After these initial observations in plants, laboratories searched for this phenomenon in other organisms. Craig C. Mello and Andrew Fire's 1998 *Nature* paper reported a potent gene silencing effect after injecting double stranded RNA into *C. elegans*. In investigating the regulation of muscle protein production, they observed that neither mRNA nor antisense RNA injections had an effect on protein production, but double-stranded RNA successfully silenced the targeted gene. As a result of this work, they coined the term *RNAi*. This discovery represented the first identification of the causative agent for the phenomenon.

Nanotechnological Approaches for Genetic Immunization

DNA vaccination is a technique for protecting an organism against disease by injecting it with genetically engineered DNA to produce an immunological response. Nucleic acid vaccines are still experimental, and have been applied to a number of viral, bacterial and parasitic models of disease, as well as to several tumour models. DNA vaccines have a number of advantages over conventional vaccines, including the ability to induce a wider range of immune response types. Many believe vaccines are among the greatest achievements of modern medicine – in industrial nations, they have eliminated naturally occurring cases of smallpox, and nearly eliminated polio, while other diseases, such as typhus, rotavirus, hepatitis A and B and others are well controlled. Conventional vaccines, however, only cover a small number of diseases, and infections that lack effective vaccines kill millions of people every year, with AIDS, hepatitis C and malaria being particularly common.

First generation vaccines are whole-organism vaccines – either live and weakened, or killed forms. Live, attenuated vaccines, such as smallpox and polio vaccines, are able to induce killer T-cell (T_C or CTL) responses, helper T-cell (T_H) responses and antibody immunity. However, there is a small risk that attenuated forms of a pathogen can revert to a dangerous form, and may still be able to cause disease in immunocompromised vaccine recipients (such as those with AIDS). While killed vaccines do not have this risk, they cannot generate specific killer T cell responses, and may not work at all for some diseases. In order to minimise these risks, so-called *second generation vaccines* were developed. These are subunit vaccines, consisting of defined protein antigens (such as tetanus or diphtheria toxoid) or recombinant protein components (such as the hepatitis B surface antigen). These, too, are able to generate T_H and antibody responses, but not killer T cell responses.

DNA vaccines are *third generation vaccines*, and are made up of a small, circular piece of bacterial DNA (called a plasmid) that has been genetically engineered to produce one or two specific proteins (antigens) from a pathogen. The vaccine DNA is injected into the cells of the body, where the "inner machinery" of the host cells "reads" the DNA and uses it to synthesize the pathogen's proteins. Because these proteins are recognised as foreign, when they are processed by the host cells and displayed on their surface, the immune system is alerted, which then triggers a range of immune responses. These DNA vaccines developed from "failed" gene therapy experiments. The first demonstration of a plasmid-induced immune response was when mice inoculated with a plasmid expressing human growth hormone elicited antibodies instead of altering growth.

Current Use

Thus far, few experimental trials have evoked a response strong enough to protect against disease, and the usefulness of the technique, while tantalizing, remains to be conclusively proven in human trials. However, in June 2006 positive results were announced for a bird flu DNA vaccine and a veterinary DNA vaccine to protect horses from West Nile virus has been approved. In August 2007, a preliminary study in DNA vaccination against multiple sclerosis was reported as being effective.

Advantages and Disadvantages

Table: Advantages And Disadvantages Of Nucleic Acid-Based Immunization

Advantages	Disadvantages
• Subunit vaccination with no risk for infection • Antigen presentation by both MHC class I and class II molecules • Able to polarise T-cell help toward type 1 or type 2 • Immune response focused only on antigen of interest • Ease of development and production[2] • Stability of vaccine for storage and shipping • Cost-effectiveness • Obviates need for peptide synthesis, expression and purification of recombinant proteins and the use of toxic adjuvants • Long-term persistence of immunogen • In vivo expression ensures protein more closely resembles normal eukaryotic structure, with accompanying post-translational modifications	• Limited to protein immunogens (not useful for non-protein based antigens such as bacterial polysaccharides) • Risk of affecting genes controlling cell growth • Possibility of inducing antibody production against DNA • Possibility of tolerance to the antigen (protein) produced • Potential for atypical processing of bacterial and parasite proteins

Plasmid Vectors for Use in Vaccination

Vector Design

DNA vaccines elicit the best immune response when highly active expression vectors are used. These are plasmids which usually consist of a strong viral promoter to drive the in vivo transcription and translation of the gene (or complementary DNA) of interest.

Intron A may sometimes be included to improve mRNA stability and hence increase protein expression. Plasmids also include a strong polyadenylation/ transcriptional termination signal, such as bovine growth hormone or rabbit beta-globulin polyadenylation sequences. Multicistronic vectors are sometimes constructed to express more than one immunogen, or to express an immunogen and an immunostimulatory protein.

Because the plasmid is the "vehicle" from which the immunogen is expressed, optimising vector design for maximal protein expression is essential. One way of enhancing protein expression is by optimising the codon usage of pathogenic mRNAs for eukaryotic cells. Pathogens often have different AT-contents than the species being immunized, so altering the gene sequence of the immunogen to reflect the codons more commonly used in the target species may improve its expression.

Another consideration is the choice of promoter. The SV40 promoter was conventionally used until research showed that vectors driven by the Rous Sarcoma Virus (RSV) promoter had much higher expression rates. More recently, expression rates have been further increased by the use of the cytomegalovirus (CMV) immediate early promoter. Inclusion of the Mason-Pfizer monkey virus (MPV)-CTE with/without rev increased envelope expression. Furthermore the CTE+rev construct was significantly more immunogenic than CTE-alone vector. Additional modifications to improve expression rates have included the insertion of enhancer sequences, synthetic introns, adenovirus tripartite leader (TPL) sequences and modifications to the polyadenylation and transcriptional termination sequences. An example of DNA vaccine plasmid is pVAC, it uses SV40 promoter.

Mechanism of Plasmids

Once the plasmid inserts itself into the nucleus of the transfected cell, it starts to encode for a gene resulting in production of peptide string of a foreign antigen. The cell on its surface displays the foreign antigen with both histocompatibility complex (MHC) classes I and class II molecule. The antigen-presenting cell then travels to the lymph nodes and presents the antigen peptide and costimulatory molecule signaled by T-cell results in initiate of the immune response.

Vaccine Insert Design

Immunogens can be targeted to various cellular compartments in order to improve antibody or cytotoxic T-cell responses. Secreted or plasma membrane-bound antigens are more effective at inducing antibody responses than cytosolic antigens, while cytotoxic T-cell responses can be improved by targeting antigens for cytoplasmic degradation and subsequent entry into the major histocompatibility complex (MHC) class I pathway. This is usually accomplished by the addition of N-terminal ubiquitin signals.

The conformation of the protein can also have an effect on antibody responses, with "ordered" structures (like viral particles) being more effective than unordered structures. Strings of minigenes (or MHC class I epitopes) from different pathogens are able to raise cytotoxic T-cell responses to a number of pathogens, especially if a TH epitope is also included.

Delivery Methods

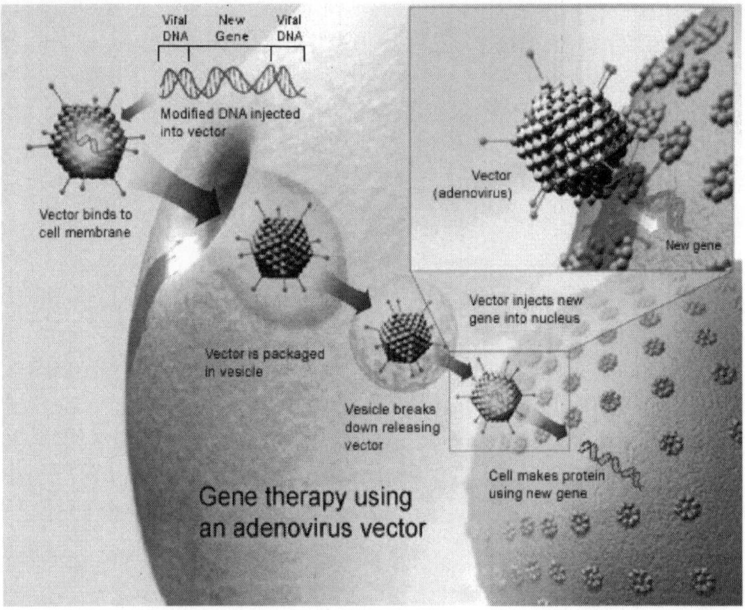

Figure: *DNA vaccine and Gene therapy techniques are similar.*

DNA vaccines have been introduced into animal tissues by a number of different methods. The two most popular approaches are injection of DNA in saline, using a standard hypodermic needle, and gene gun delivery. A schematic outline of the construction of a DNA vaccine plasmid and its subsequent delivery by these two methods into a host is illustrated at Scientific American. Injection in saline is normally conducted intramuscularly (IM) in skeletal muscle, or intradermally (ID), with DNA being delivered to the extracellular spaces. This can be assisted by electroporation; by temporarily damaging muscle fibres with myotoxins such as bupivacaine; or by using hypertonic solutions of saline or sucrose. Immune responses to this method of delivery can be affected by many factors, including needle type, needle alignment, speed of injection, volume of injection, muscle type, and age, sex and physiological condition of the animal being injected.

Gene gun delivery, the other commonly used method of delivery, ballistically accelerates plasmid DNA (pDNA) that has been adsorbed onto gold or tungsten microparticles into the target cells, using compressed helium as an accelerant. Alternative delivery methods have included aerosol instillation of naked DNA on mucosal surfaces, such as the nasal and lung mucosa, and topical administration of pDNA to the eye and vaginal mucosa. Mucosal surface delivery has also been achieved using cationic liposome-DNA preparations, biodegradable microspheres, attenuated *Shigella* or *Listeria* vectors for oral administration to the intestinal mucosa, and recombinant adenovirus vectors.

The method of delivery determines the dose of DNA required to raise an effective immune response. Saline injections require variable amounts of DNA, from 10 ìg-1 mg, whereas gene gun deliveries require 100 to 1000 times less DNA than intramuscular saline injection to raise an effective immune response. Generally, 0.2 ìg – 20 ìg are required, although quantities as low as 16 ng have been reported. These quantities vary from species to species, with mice, for example, requiring approximately 10 times less DNA than primates. Saline injections require more DNA because the DNA is delivered to the extracellular spaces of the target tissue (normally muscle), where it has to overcome physical barriers (such as the basal lamina and large amounts of connective tissue, to mention a few) before it is taken up by the cells, while gene gun deliveries bombard DNA directly into the cells, resulting in less "wastage".

Table: *Summary of Plasmid DNA delivery methods*

Method of Delivery	***Formulation of DNA***	***Target Tissue***	***Amount of DNA***	
Parenteral	Injection (hypodermic needle)	Aqueous solution in saline	IM (skeletal); ID; (IV, subcutaneous and intraperitoneal with variable success)	Large amounts (approximately 100-200 µg)
	Gene Gun	DNA-coated gold beads	ED (abdominal skin); vaginal mucosa; surgically exposed muscle and other organs	Small amounts (as little as 16 ng)
	Pneumatic (Jet) Injection	Aqueous solution	ED	Very high (as much as 300 µg)
Topical application	Aqueous solution	Ocular; intravaginal	Small amounts (up to 100 µg)	
Cytofectin-mediated	Liposomes (cationic); microspheres; recombinant adenovirus vectors; attenuated Shigella vector; aerosolised cationic lipid formulations	IM; IV (to transfect tissues systemically); intraperitoneal; oral immunization to the intestinal mucosa; nasal/lung mucosal membranes	variable	

Table: *Advantages and disadvantages of commonly used DNA vaccine delivery methods*

Another approach to DNA vaccination is expression library immunization (ELI). Using this technique, potentially all the genes from a pathogen can be delivered at one time, which may be useful for pathogens which are difficult to attenuate or culture. ELI can be used to identify which of the pathogen's genes induce a protective response.

This has been tested with *Mycoplasma pulmonis*, a murine lung pathogen with a relatively small genome, and it was found that even partial expression libraries can induce protection from subsequent challenge.

Method of Delivery	***Advantage***	***Disadvantage***
Intramuscular or Intradermal injection	• No special delivery mechanism • Permanent or semi-permanent expression • pDNA spreads rapidly throughout the body	• Inefficient site for uptake due to morphology of muscle tissue • Relatively large amounts of DNA used • Th1 response may not be the response required
Gene Gun	• DNA bombarded directly into cells • Small amounts DNA	• Th2 response may not be the response required • Requires inert particles as carrier
Jet injection	• No particles required • DNA can be delivered to cells mm to cm below skin surface	• Significant shearing of DNA after high-pressure expulsion • 10-fold lower expression, and lower immune response • Requires large amounts of DNA (up to 300 µg)
Liposome-mediated delivery	• High levels of immune response can be generated • Can increase transfection of intravenously delivered pDNA • Intravenously delivered liposome-DNA complexes can potentially transfect all tissues • Intranasally delivered liposome-DNA complexes can result in expression in distal mucosa as well as nasal muscosa and the generation of IgA antibodies	• Toxicity • Ineffectiveness in serum • Risk of disease or immune reactions

Immune Response Raised by DNA Vaccines

Helper T-Cell Responses

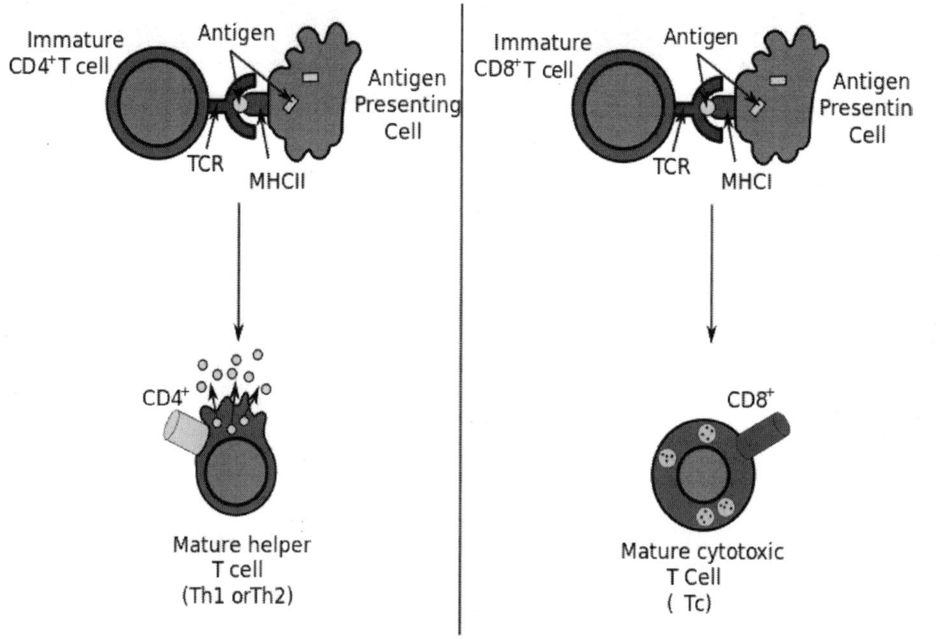

Figure: *Antigen presentation stimulates T cells to become either "cytotoxic" CD8+ cells or "helper" CD4+ cells. Cytotoxic cells directly attack other cells carrying certain foreign or abnormal molecules on their surfaces. Helper T cells, or Th cells, coordinate immune responses by communicating with other cells. In most cases, T cells only recognise an antigen if it is carried on the surface of a cell by one of the body's own MHC, or major histocompatibility complex, molecules.*

DNA immunization is able to raise a range of T_H responses, including lymphoproliferation and the generation of a variety of cytokine profiles. A major advantage of DNA vaccines is the ease with which they can be manipulated to bias the type of T-cell help towards a TH1 or TH2 response. Each type of response has distinctive patterns of lymphokine and chemokine expression, specific types of immunoglobulins expressed, patterns of lymphocyte trafficking, and types of innate immune responses generated.

Raising of Different Types of T-Cell Help

The type of T-cell help raised is influenced by the method of delivery and the type of immunogen expressed, as well as the targeting of different lymphoid compartments. Generally, saline needle injections (either IM or ID) tend to induce TH1 responses, while gene gun delivery raises TH2 responses. This is true for intracellular and plasma membrane-bound antigens, but not for secreted antigens, which seem to generate TH2 responses, regardless of the method of delivery.

Generally the type of T-cell help raised is stable over time, and does not change when challenged or after subsequent immunizations which would normally have raised the opposite type of response in a naïve animal. However, Mor *et al.*. (1995) immunized and boosted mice with pDNA encoding the circumsporozoite protein of the mouse malarial parasite *Plasmodium yoelii* (PyCSP) and found that the initial TH2 response changed, after boosting, to a TH1 response.

Mechanistic Basis for Different Types of T-Cell Help

It is not understood how these different methods of DNA immunization, or the forms of antigen expressed, raise different profiles of T-cell help. It was thought that the relatively large amounts of DNA used in IM injection were responsible for the induction of TH1 responses. However, evidence has shown no differences in TH type due to dose. It has been postulated that the type of T-cell help raised is determined by the differentiated state of antigen presenting cells. Dendritic cells can differentiate to secrete IL-12 (which supports TH1 cell development) or IL-4 (which supports TH2 responses). pDNA injected by needle is endocytosed into the dendritic cell, which is then stimulated to differentiate for TH1 cytokine production, while the gene gun bombards the DNA directly into the cell, thus bypassing TH1 stimulation.

Practical Uses of Polarised T-Cell Help

This polarisation in T-cell help is useful in influencing allergic responses and autoimmune diseases. In autoimmune diseases, the goal would be to shift the self-destructive TH1 response (with its associated cytotoxic T cell activity) to a non-destructive TH2 response. This has been successfully applied in predisease priming for the desired type of response in preclinical models and somewhat successful in shifting the response for an already established disease.

Cytotoxic T-Cell Responses

One of the greatest advantages of DNA vaccines is that they are able to induce cytotoxic T lymphocytes (CTL) without the inherent risk associated with live vaccines. CTL responses can be raised against immunodominant and immunorecessive CTL epitopes, as well as subdominant CTL epitopes, in a manner which appears to mimic natural infection. This may prove to be a useful tool in assessing CTL epitopes of an antigen, and their role in providing immunity.

Cytotoxic T-cells recognise small peptides (8-10 amino acids) complexed to MHC class I molecules (Restifo et al., 1995). These peptides are derived from endogenous cytosolic proteins which are degraded and delivered to the nascent MHC class I molecule within the endoplasmic reticulum (ER). Targeting gene products directly to the ER (by the addition of an amino-terminal insertion sequence) should thus enhance CTL responses. This has been successfully demonstrated using recombinant vaccinia viruses expressing influenza proteins, but the principle should be applicable to DNA vaccines too. Targeting antigens for intracellular degradation (and thus entry into the MHC class I pathway) by the addition of ubiquitin signal sequences, or mutation of other signal sequences, has also been shown to be effective at increasing CTL responses.

CTL responses can also be enhanced by co-inoculation with co-stimulatory molecules such as B7-1 or B7-2 for DNA vaccines against influenza nucleoprotein, or GM-CSF for DNA vaccines against the murine malaria model *P. yoelii*. Co-inoculation with plasmids encoding co-stimulatory molecules IL-12 and TCA3 have also been shown to increase CTL activity against HIV-1 and influenza nucleoprotein antigens.

Humoral (Antibody) Response

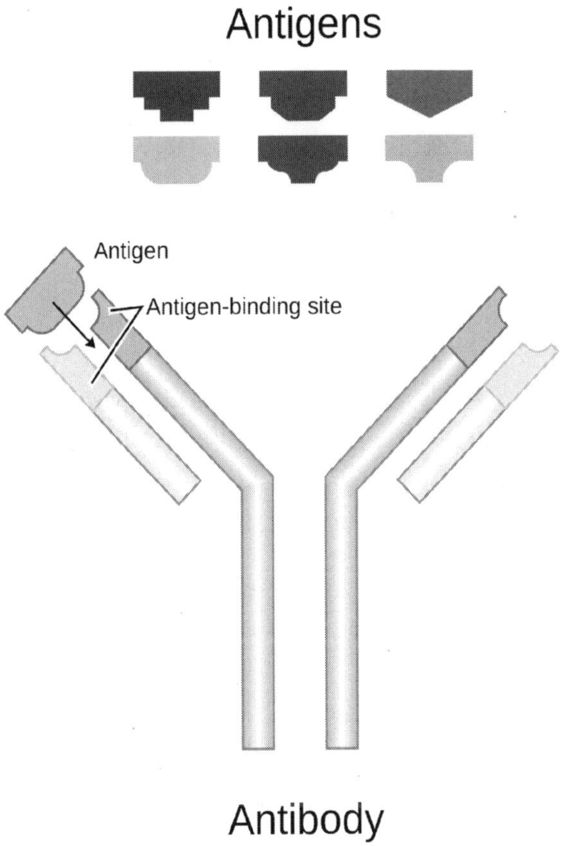

Figure: *Schematic diagram of an antibody and antigens*

Antibody responses elicited by DNA vaccinations are influenced by a number of variables, including type of antigen encoded; location of expressed antigen (i.e. intracellular vs. secreted); number, frequency and dose of immunizations; site and method of antigen delivery, to name a few.

Kinetics of Antibody Response

Humoral responses after a single DNA injection can be much longer-lived than after a single injection with a recombinant protein. Antibody responses against hepatitis B virus (HBV) envelope protein (HBsAg) have been sustained for up to 74 weeks without boost, while lifelong maintenance of protective response to influenza haemagglutinin has been demonstrated in mice after gene gun delivery. Antibody-secreting cells migrate to the bone marrow and spleen for long-term antibody production, and are generally localised there after one year.

Comparisons of antibody responses generated by natural (viral) infection, immunization with recombinant protein and immunization with pDNA are summarised in Table 4. DNA-raised antibody responses rise much more slowly than when natural infection or recombinant protein immunization occurs. It can take as long as 12 weeks to reach peak titres in mice, although boosting can increase the rate of antibody production. This slow response is probably due to the low levels of antigen expressed over several weeks, which supports both primary and secondary phases of antibody response. DNA vaccine expressing HBV small and middle envelope protein was injected into adults with chronic hepatitis. The vaccine resulted in specific interferon gamma cell production. Also specific T-cells for middle envelop proteins antigens were developed. The immune response of the patients was not robust enough to control HBV infection (Mancini - Bourgine et al.)

Table: *Comparison of T-Dependent Antibody Responses raise by DNA Immunizations, Protein Inoculations and Viral Infections*

Method of Immunization			
DNA Vaccine	***Recombinant protein***	***Natural Infection***	
Amount of inducing antigen	ng	μg	? (ng-μg)
Duration of Ag presentation	several weeks	< 1 week	several weeks
Kinetics of Ab response	slow rise	rapid rise	rapid rise
Number of inoculations to obtain high avidity IgG and migration of ASC to bone marrow	one	two	one
Ab isotype (murine models)	C'-dependent or C'-independent	C'-dependent	C'-independent

Additionally, the titres of specific antibodies raised by DNA vaccination are lower than those obtained after vaccination with a recombinant protein. However, DNA immunization-induced antibodies show greater affinity to native epitopes than recombinant protein-induced antibodies. In other words, DNA immunization induces a qualitatively superior response. Antibody can be induced after just one vaccination with DNA, whereas recombinant protein vaccinations generally require a boost. As mentioned previously, DNA immunization can be used to bias the TH profile of the immune response, and thus the antibody isotype, which is not possible with either natural infection or recombinant protein immunization. Antibody responses generated by DNA are useful not just in vaccination but as a preparative tool, too. For example, polyclonal and monoclonal antibodies can be generated for use as reagents.

Mechanistic Basis for DNA Raised Immune Responses

DNA Uptake Mechanism

When DNA uptake and subsequent expression was first demonstrated *in vivo* in muscle cells, it was thought that these cells were unique in this ability because of their extensive network of T-tubules. Using electron microscopy, it was proposed that DNA uptake was facilitated by caveolae (or, non-clathrin coated pits). However, subsequent research revealed that other cells (such as keratinocytes, fibroblasts and epithelial Langerhans cells) could also internalize DNA. This phenomenon has not been the subject of much research, so the actual mechanism of DNA uptake is not known.

Two theories are currently popular – that *in vivo* uptake of DNA occurs non-specifically, in a method similar to phago- or pinocytosis, or through specific receptors. These might include a 30kDa surface receptor, or macrophage scavenger receptors. The 30kDa surface receptor binds very specifically to 4500-bp genomic DNA fragments (which are then internalised) and is found on professional APCs and T-cells. Macrophage scavenger receptors bind to a variety of macromolecules, including polyribonucleotides, and are thus also candidates for DNA uptake. Receptor mediated DNA uptake could be facilitated by the presence of polyguanylate sequences. Further research into this mechanism might seem pointless, considering that gene gun delivery systems, cationic liposome packaging, and other delivery methods bypass this entry method, but understanding it might be useful in reducing costs (e.g. by reducing the requirement for cytofectins), which will be important in the food animals industry.

Antigen Presentation by Bone Marrow-Derived Cells

Studies using chimeric mice have shown that antigen is presented by bone-marrow derived cells, which include dendritic cells, macrophages and specialised B-cells called professional antigen presenting cells (APC) Iwasaki et al., 1997). After gene gun inoculation to the skin, transfected Langerhans cells migrate to the draining lymph node to present antigen. After IM and ID injections, dendritic cells have also been found to present antigen in the draining lymph node and transfected macrophages have been found in the peripheral blood.

Besides direct transfection of dendritic cells or macrophages, cross priming is also known to occur following IM, ID and gene gun deliveries of DNA. Cross priming occurs when a bone marrow-derived cell presents peptides from proteins synthesised in another cell in the context of MHC class 1. This can prime cytotoxic T-cell responses and seems to be important for a full primary immune response.

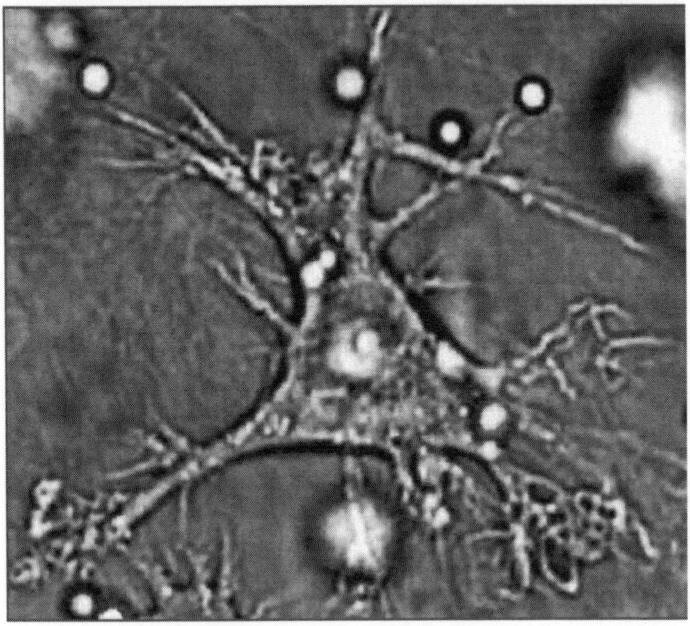

Figure: *A dendritic cell.*

Role of the Target Site

IM and ID delivery of DNA initiate immune responses differently. In the skin, keratinocytes, fibroblasts and Langerhans cells take up and express antigen, and are responsible for inducing a primary antibody response. Transfected Langerhans cells migrate out of the skin (within 12 hours) to the draining lymph node where they prime secondary B- and T-cell responses. In skeletal muscle, on the other hand, striated muscle cells are most frequently transfected, but seem to be unimportant in mounting an immune response. Instead, IM inoculated DNA "washes" into the draining lymph node within minutes, where distal dendritic cells are transfected and then initiate an immune response. Transfected myocytes seem to act as a "reservoir" of antigen for trafficking professional APCs.

Maintenance of Immune Response

DNA vaccination generates an effective immune memory via the display of antigen-antibody complexes on follicular dendritic cells (FDC), which are potent B-cell stimulators. T-cells can be stimulated by similar, germinal centre dendritic cells. FDC are able to generate an immune memory because antibodies production "overlaps" long-term expression of antigen, allowing antigen-antibody immunocomplexes to form and be displayed by FDC.

Interferons

Both helper and cytotoxic T-cells can control viral infections by secreting interferons. Cytotoxic T cells usually kill virally infected cells. However, they can also be stimulated to secrete antiviral cytokines such as IFN-ã and TNF-á, which don't kill the cell but place severe limitations on viral infection by down-regulating the expression of viral components. DNA vaccinations can thus be used to curb viral infections by non-destructive IFN-mediated control. This has been demonstrated for the hepatitis B virus. IFN-ã is also critically important in controlling malaria infections, and should be taken into consideration when developing anti-malarial DNA vaccines.

Modulation of the Immune Response

Cytokine Modulation

For a vaccine to be effective, it must induce an appropriate immune response for a given pathogen, and the ability of DNA vaccines to polarise T-cell help towards TH1 or TH2 profiles, and generate CTL and/or antibody when required, is a great advantage in this regard. This can be accomplished by modifications to the form of antigen expressed (i.e. intracellular vs. secreted), the method and route of delivery, and the dose of DNA delivered. However, it can also be accomplished by the co-administration of plasmid DNA encoding immune regulatory molecules, i.e. cytokines, lymphokines or co-stimulatory molecules. These "genetic adjuvants" can be administered a number of ways:

- as a mixture of 2 separate plasmids, one encoding the immunogen and the other encoding the cytokine;
- as a single bi- or polycistronic vector, separated by spacer regions; or
- as a plasmid-encoded chimera, or fusion protein.

In general, co-administration of pro-inflammatory agents (such as various interleukins, tumor necrosis factor, and GM-CSF) plus TH2 inducing cytokines increase antibody responses, whereas pro-inflammatory agents and TH1 inducing cytokines decrease humoral responses and increase cytotoxic responses (which is more important in viral protection, for example). Co-stimulatory molecules like B7-1, B7-2 and CD40L are also sometimes used.

This concept has been successfully applied in topical administration of pDNA encoding IL-10. Plasmid encoding B7-1 (a ligand on APCs) has successfully enhanced the immune response in anti-tumour models, and mixing plasmids encoding GM-CSF and the circumsporozoite protein of P. yoelii (PyCSP) has enhanced protection against subsequent challenge (whereas plasmid-encoded PyCSP alone did not). It was proposed that GM-CSF may cause dendritic cells to present antigen more efficiently, and enhance IL-2 production and TH cell activation, thus driving the increased immune response.

This can be further enhanced by first priming with a pPyCSP and pGM-CSF mixture, and later boosting with a recombinant poxvirus expressing PyCSP. However, co-injection of plasmids encoding GM-CSF (or IFN-ã, or IL-2) and a fusion protein of *P. chabaudi* merozoite surface protein 1 (C-terminus)-hepatitis B virus surface protein (PcMSP1-HBs) actually abolished protection against challenge, compared to protection acquired by delivery of pPcMSP1-HBs alone.

The advantages of using genetic adjuvants are their low cost and simplicity of administration, as well as avoidance of unstable recombinant cytokines and potentially toxic, "conventional" adjuvants (such as alum, calcium phosphate, monophosphoryl lipid A, cholera toxin, cationic and mannan-coated liposomes, QS21, carboxymethylcellulose and ubenimix). However, the potential toxicity of prolonged cytokine expression has not been established, and in many commercially important animal species, cytokine genes still need to be identified and isolated. In addition, various plasmid encoded cytokines modulate the immune system differently according to the time of delivery. For example, some cytokine plasmid DNAs are best delivered after the immunogen pDNA, because pre- or co-delivery can actually decrease specific responses, and increase non-specific responses.

Immunostimulatory CpG Motifs

Plasmid DNA itself appears to have an adjuvant effect on the immune system. Bacterially derived DNA has been found to trigger innate immune defence mechanisms, the activation of dendritic cells, and the production of TH1 cytokines. This is due to recognition of certain CpG dinucleotide sequences which are immunostimulatory. CpG stimulatory (CpG-S) sequences occur twenty times more frequently in bacterially derived DNA than in eukaryotes. This is because eukaryotes exhibit "CpG suppression" – i.e. CpG dinucleotide pairs occur much less frequently than expected. Additionally, CpG-S sequences are hypomethylated. This occurs frequently in bacterial DNA, while CpG motifs occurring in eukaryotes are all methylated at the cytosine nucleotide. In contrast, nucleotide sequences which inhibit the activation of an immune response (termed CpG neutralising, or CpG-N) are over represented in eukaryotic genomes. The optimal immunostimulatory sequence has been found to be an unmethylated CpG dinucleotide flanked by two 5' purines and two 3' pyrimidines. Additionally, flanking regions outside this immunostimulatory hexamer must be guanine-rich to ensure binding and uptake into target cells.

The innate system works synergistically with the adaptive immune system to mount a response against the DNA encoded protein. CpG-S sequences induce polyclonal B-cell activation and the upregulation of cytokine expression and secretion. Stimulated macrophages secrete IL-12, IL-18, TNF-α, IFN-α, IFN-β and IFN-λ, while stimulated B-cells secrete IL-6 and some IL-12.

Manipulation of CpG-S and CpG-N sequences in the plasmid backbone of DNA vaccines can ensure the success of the immune response to the encoded antigen, and drive the immune response toward a TH1 phenotype. This is useful if a pathogen requires a TH response for protection. CpG-S sequences have also been used as external adjuvants for both DNA and recombinant protein vaccination with variable success rates. Other organisms with hypomethylated CpG motifs have also demonstrated the stimulation of polyclonal B-cell expansion. However, the mechanism behind this may be more complicated than simple methylation – hypomethylated murine DNA has not been found to mount an immune response.

Most of the evidence for the existence of immunostimulatory CpG sequences comes from murine studies. Clearly, extrapolation of this data to other species should be done with caution – different species may require different flanking sequences, as binding specificities of scavenger receptors differ between species. Additionally, species such as ruminants may be insensitive to immunostimulatory sequences due to the large gastrointestinal load they exhibit. Further research may be useful in the optimisation of DNA vaccination, especially in the food animal industry.

Alternative Boosts

DNA-primed immune responses can be boosted by the administration of recombinant protein or recombinant poxviruses. "Prime-boost" strategies with recombinant protein have successfully increased both neutralising antibody titre, and antibody avidity and persistence, for weak immunogens, such as HIV-1 envelope protein. Recombinant virus boosts have been shown to be very efficient at boosting DNA-primed CTL responses. Priming with DNA focuses the immune response on the required immunogen, while boosting with the recombinant virus provides a larger amount of expressed antigen, leading to a large increase in specific CTL responses.

Prime-boost strategies have been successful in inducing protection against malarial challenge in a number of studies. Primed mice with plasmid DNA encoding *Plasmodium yoelii* circumsporozoite surface protein (PyCSP), then boosted with a recombinant vaccinia virus expressing the same protein had significantly higher levels of antibody, CTL activity and IFN-ã, and hence higher levels of protection, than mice immunized and boosted with plasmid DNA alone. This can be further enhanced by priming with a mixture of plasmids encoding PyCSP and murine GM-CSF, before boosting with recombinant vaccinia virus. An effective prime-boost strategy for the simian malarial model *P. knowlesi* has also been demonstrated. Rhesus monkeys were primed with a multicomponent, multistage DNA vaccine encoding two liver-stage antigens - the circumsporozoite surface protein (PkCSP) and sporozoite surface protein 2 (PkSSP2) - and two blood stage antigens - the apical merozoite surface protein 1 (PkAMA1) and merozoite surface protein 1 (PkMSP1p42). They were then boosted with a recombinant canarypox virus encoding all four antigens (ALVAC-4). Immunized monkeys developed antibodies against sporozoites and infected erythrocytes, and IFN-ã-secreting T-cell responses against peptides from PkCSP. Partial protection against sporozoite challenge was achieved, and mean parasitemia was significantly reduced, compared to control monkeys. These models, while not ideal for extrapolation to *P. falciparum* in humans, will be important in pre-clinical trials.

Additional Methods of Enhancing DNA-Raised Immune Responses

Formulations of DNA

The efficiency of DNA immunization can be improved by stabilising DNA against degradation, and increasing the efficiency of delivery of DNA into antigen presenting cells. This has been demonstrated by coating biodegradable cationic microparticles (such as poly(lactide-co-glycolide) formulated with cetyltrimethylammonium bromide) with DNA. Such DNA-coated microparticles can be as effective at raising CTL as recombinant vaccinia viruses, especially when mixed with alum. Particles 300 nm in diameter appear to be most efficient for uptake by antigen presenting cells.

Alphavirus Vectors

Recombinant alphavirus-based vectors have also been used to improve DNA vaccination efficiency. The gene encoding the antigen of interest is inserted into the alphavirus replicon, replacing structural genes but leaving non-structural replicase genes intact. The Sindbis virus and Semliki Forest virus have been used to build recombinant alphavirus replicons. Unlike conventional DNA vaccinations, however, alphavirus vectors kill transfected cells, and are only transiently expressed. Also, alphavirus replicase genes are expressed in addition to the vaccine insert. It is not clear how alphavirus replicons raise an immune response, but it is thought that this may be due to the high levels of protein expressed by this vector, replicon-induced cytokine responses, or replicon-induced apoptosis leading to enhanced antigen uptake by dendritic cells.

5

DNA Functionalized Gold Nanoparticles for Metabolite and Nucleic Acid Detection

Spherical nucleic acids (SNAs) – defined as structures which are an arrangement of densely packed, highly oriented nucleic acids in a spherical geometry – were first introduced in 1996 by the Mirkin group at Northwestern University. The unique arrangement and orientation of one-dimensional linear nucleic acids within this three-dimensional framework results in fundamentally new chemical, biological, and physical properties, which represent a paradigm shift in the use of nucleic acids for intracellular gene regulation, molecular diagnostics, and materials synthesis applications.

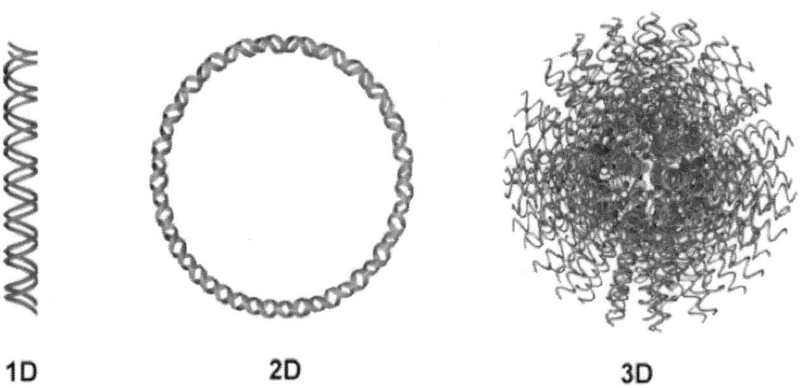

Figure: *Three important classes of nucleic acids: one-dimensional linear, two-dimensional circular, and three-dimensional spherical.*

Structure and Function

The first SNA consisted of 3' alkanethiol-terminated, single-stranded oligonucleotides (short DNA sequences) covalently attached to the surface of spherical gold nanoparticles. A dense nucleic acid loading was achieved through a series of salt additions, in which positively charged counterions were used to reduce electrostatic repulsion between adjacent negatively charged DNA strands and thereby enable more efficient DNA packing onto the nanoparticle surface.

One- and two-dimensional forms of nucleic acids (e.g., single strands, linear duplexes, and plasmids) are important biological machinery for the storage and transmission of genetic information. Underlying this utility are the bioprogrammable interactions between complementary nucleotide bases. Scientists and engineers have been synthesizing and, in certain cases, mass-producing such structures for decades in an effort to understand and exploit this elegant recognition motif. The recognition abilities of nucleic acids are enhanced when arranged in a spherical geometry, which allows for polyvalent interactions. This polyvalency, along with the high density and degree of orientation described above, helps to explain why a three-dimensional nucleic acid structure fundamentally composed of linear, one-dimensional nucleic acids exhibits dramatically different properties than its lower-dimensional constituents.

Property	Spherical Nucleic Acids	Linear Nucleic Acids
Melting Transition	Cooperative and Narrow (~2-8 °C)	Broad (~20 °C)
Cellular Uptake	Transfection agents NOT required, 1-1.5 x 10^6 NPs per cell†	Transfection agents required (e.g. DharmaFECT™, Lipofectamine™, Ca^{2+})
Immune Response	Minimal	Elevated interferon-β levels (25-fold increase compared to DNA-AuNP conjugates)
Stability	Nuclease resistance due to high local salt concentration	Subject to degradation by nucleases (e.g. DNAse degradation 4x higher rate than SNA)
Properties from Inorganic Core	Plasmonic, Catalytic, Magnetic, Luminescent	n/a
Binding Strength‡	K_{eq} = 1.8 x 10^{14} Activated binding motifs	K_{eq} = 1.8 x 10^{12}

†Numbers vary depending on cell type and nucleic acid sequence
‡K_{eq} values for 15-mer AT-rich strand

Figure: *Properties of spherical nucleic acids (SNAs) versus linear nucleic acids.*

Two decades of research on SNAs have revealed that they exhibit a synergistic combination of properties resulting from the inorganic core and the nucleic acid shell. The inorganic nanoparticle core (in addition to gold, silver, iron oxide, silica, and semiconductor materials have also been used) serves two purposes:

1) it imparts upon the conjugate novel physical and chemical properties (e.g. plasmonic, catalytic, magnetic, luminescent), and

2) it acts as a scaffold for the assembly and orientation of nucleic acids into the dense arrangement that gives rise to many of their functional properties.

The nucleic acid shell gives the conjugate unique, programmable chemical and biological recognition abilities that offer greater binding strengths and higher duplex stabilities compared to the same sequence of linear DNA, cooperative melting behaviour with DNA strands of a complementary sequence, and enhanced cellular uptake without the use of transfection agents.

Recent studies have shown that one can crosslink the DNA strands at their base, and subsequently dissolve the inorganic core with KCN or I_2 to create a new coreless form of SNA, which exhibits many of the hallmark properties of the original DNA-nanoparticle conjugate. This work underscores one of the fundamental features of SNAs: the properties of these nanomaterials are core-independent, derived from the orientation and packing of the nucleic acid shell.

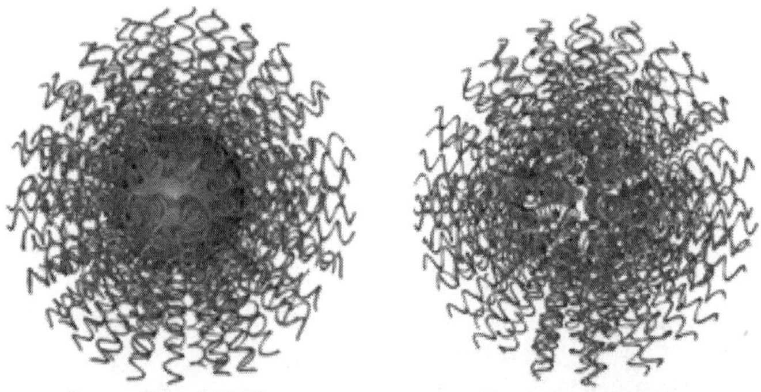

Figure: *Gold nanoparticle filled and core-less spherical nucleic acid structures (SNAs).*

Due to their unique structure and function, SNAs occupy a materials space distinct from "DNA nanotechnology and origami." With DNA origami, such structures are synthesized via DNA hybridization events. In contrast, the SNA structure can be synthesized independent of nucleic acid sequence and hybridization, instead relying upon robust chemical bond formation between nanoparticles and nucleic acid adsorbates.

Furthermore, DNA origami uses DNA hybridization interactions to realise a final structure, whereas SNAs and other forms of three-dimensional nucleic acids (structures templated with a triangular prism, rod, octahedra, or rhombic dodecadhedra-shaped nanoparticles) utilise the nanoparticle core to arrange the linear nucleic acid components into functional forms. Indeed, it is the particle core that dictates the shape of the SNA, and to date, single-stranded and double-stranded versions of these materials have been created that consist of DNA, LNA, and RNA.

SNAs should also not be confused with their monovalent analogues – individual particles coupled to a single DNA strand. Such single strand-nanoparticle conjugate structures have led to interesting advances in their own right, but do not exhibit the unique properties of SNAs.

Applications and Societal Benefit

Intracellular Gene Regulation

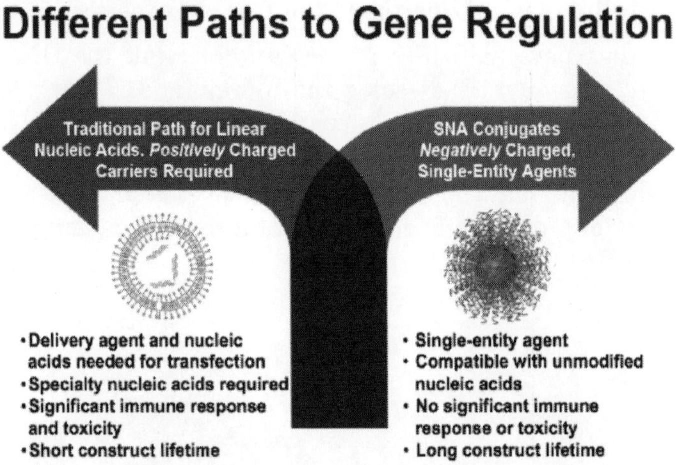

Figure: *Nucleic acids arranged in a spherical geometry offer a fundamentally new path toward gene regulation. Benefits to this approach include the ability to enter cells without precomplexation with a transfection agent, nuclease resistance, and minimal immune response.*

SNAs also have advanced the field of nanotherapeutics as extraordinarily useful gene regulation agents. Despite their high negative charge, these nanostructures are naturally taken up by cells, without the need for positively charged polymeric co-carriers, and are highly effective as gene regulation agents in both siRNA and antisense gene regulation pathways. This discovery represents a paradigm shift for the field, since previous lore predicted that nucleic acids would not efficiently enter cells without positively charged co-carriers (Fig. 4). Although this is true for linear nucleic acids, it is not the case for SNAs. This stunning result is because SNAs, unlike their linear counterparts, have the ability to complex scavenger proteins and thereby facilitate endocytosis. Investigation of the mechanism for SNA uptake revealed that these nanostructures exhibit an ability to deactivate enzymes responsible for triggering cellular immune response and oligonucleotide degradation. These materials constitute a powerful single-entity gene regulation platform for intracellular biochemical and biological applications and represent therapeutic lead structures for a wide variety of oncology treatments. Indeed, the first therapeutic versions of these structures are currently entering clinical trials

SNAs packed gold nanoparticles are able to deliver small interfering RNA (siRNA) to treat glioblastoma multiforme in a proof-of-concept study using mouse model, reported by the research team led by Mirkin. The iRNA targets Bcl2Like12 gene overexpressed in glioblastoma tumors and silences the oncogene. The gold nanoparticles injected intravenously cross the blood-brain barrier and naturally find their target in the brain. In the animal model, the treatment caused a 20% increase in survival rate and 3 to 4 fold reduction in tumor size. Mirkin indicated that this novel therapeutic approach establishes a platform for treating a wide range of diseases.

Molecular Diagnostics

Figure: *The FDA-cleared Verigene System, developed and commercialized by Nanosphere, Inc., a company spun-out of research projects initiated in Mirkin's laboratory at Northwestern University. There are nine FDA-cleared panel assays based upon SNAs, now used by hospitals all over the world.*

Spherical nucleic acids have catalyzed world-wide interest in using well-characterized nanostructures as novel labels for in vitro biodetection schemes and intracellular assays, and as potent cell transfection, therapeutic, and gene regulation materials. SNAs have enabled the first commercialized molecular medical diagnostic systems of the modern era of nanotechnology (Fig. 5). The FDA-cleared Verigene System, commercialized by Nanosphere, is now sold in over twenty countries. This technology allows the detection of markers for many diseases, including infectious disease and cancers, with a sensitivity and selectivity far exceeding that of conventional diagnostic tools. Indeed, the Verigene is transforming patient care by transitioning molecular diagnostic screening from centralized, often remote, analytical laboratories to the local hospital setting, which dramatically decreases the time required for diagnosis. Further, the Verigene has enabled the identification of new markers for Alzheimer's disease, HIV, and cardiac disease as well as new tests for the early detection of a variety of forms of prostate cancer. These medical diagnostic and therapeutic tools have already saved or improved many lives. As such, SNAs represent an important class of diagnostic probes that will enable fundamental discoveries and empower physicians to make quick and accurate decisions about patient care regarding almost any disease with a genetic basis.

Intracellular Probes

NanoFlares utilise the SNA architecture for intracellular mRNA detection. In this design, alkanethiol-terminated antisense DNA strands (complementary to a target mRNA strand within cells) are attached to the surface of a gold nanoparticle. Fluorophore-labelled "reporter strands" are then hybridized to the SNA construct to form the NanoFlare. When the fluorophore labels are brought within close proximity of the gold surface, as controlled by programmable nucleic acid hybridization, their fluorescence is quenched (Fig. 6). Following cellular entry (enabled by the SNA geometry), reporter strands are concomitantly driven from the NanoFlare and replaced by the longer, target mRNA sequence. Note that mRNA binding is thermodynamically favoured since the strands holding the reporter sequence have greater overlap of their nucleotide sequence with the target mRNA. Upon reporter strand release, dye fluorescence is no longer quenched by the gold nanoparticle core, and increased fluorescence is observed. This method for RNA detection provides the only way to sort live cells based upon genetic content, and has thus enabled significant advances in disease diagnosis and cellular genetics. AuraSense and AuraSense Therapeutics are two start-up companies founded to advance these SNA structures in several sectors of the biomedical industry. In 2011, AuraSense entered into partnership with EMD-Millipore to commercialize NanoFlares under the SmartFlare trade name.

Materials Synthesis

SNAs also have been utilised to develop an entirely new field of materials science – one that focuses on using SNAs as synthetically-programmable building blocks in the construction of DNA-nanoparticle superlattices (Fig. 7). In 2011, a landmark paper was published in Science that defines a set of design rules for making superlattice structures of tailorable crystallographic symmetry and lattice parameters with sub-nm precision. The rules are described as follows:

1. When all DNA-NPs in a system possess equal hydrodynamic radii, each NP in the thermodynamic product will maximize the number of nearest neighbours to which it can form DNA connections.
2. When two lattices are of similar stability, the kinetic product can be produced by slowing the rate at which individual DNA linkers dehybridize and subsequently rehybridize.

3. The overall hydrodynamic radius of a DNA-NP, rather than the sizes of its individual NP or oligonucleotide components, dicates its assembly and packing behaviour.
4. In a binary system, the size ratio and DNA linker ratio between two particles dictate the thermodynamically favoured crystal structure.
5. Two systems with the same size ratio and DNA linker ratio exhibit the same thermodynamic product.
6. The most stable crystal structure will maximize all possible types of DNA sequence-specific hybridization interactions.

These design rules are analogous to Pauling's rules for ionic crystals, but ultimately more powerful. For example, when using atomic or ionic building blocks in the construction of materials, the crystal structure, symmetry, and spacing are fixed by atomic radii and electronegativity. However, in the nanoparticle-based system, crystal structure can be tuned independent of the nanoparticle size and composition by simply adjusting the length and sequence of the attached DNA. This has enabled the construction of novel crystal structures for several materials systems and even crystal structures with no mineral equivalents. The ability to place nanoparticles of any composition and shape at any location in a well-defined crystalline lattice with nm-scale precision should have far-reaching implications in areas ranging from catalysis to photonics to energy.

Economic Impact

The economic impact of SNA technology is substantial and rapidly growing. Three companies have been founded that are based on SNA technology - Nanosphere, AuraSense, and AuraSense Therapeutics (founded in 2000, 2009, and 2011, respectively). These companies have employed hundreds of people and commercialized over 1300 products. Nanosphere was one of the first nanotechnology-based biotechnology firms to go public in late 2007. The FDA-cleared Verigene System has been commercialized by Nanosphere, and nine accompanying FDA-cleared panel assays (Warfarin Metabolism, F5/F2/MTHFR, Influenza A and B, Staphylococcus aureus, 2009 H1N1) are now used by hospitals all over the world. Further, hundreds of research laboratories are currently utilising these structures in all sorts of applications spanning genomics and proteomics.

DNA Functionalized Gold Nanoparticles for Bioanalysis

Gold nanoparticles (Au NPs) have become one of the most interesting sensing materials because of their unique size- and shape-dependent optical properties, high extinction coefficients, and super-quenching capability. Au NPs that are bioconjugated with DNA (DNA-Au NPs) have been demonstrated for selective and sensitive detection of analytes such as mercury(II) ions, platelet-derived growth factor (PDGF), and adenosine triphosphate (ATP).

The last 16 years have witnessed the landmark development of polyvalent thiolated DNA functionalized gold nanoparticles (AuNPs) possessing striking properties within the emerging field of nanobiotechnology. Many novel properties of this hybrid nanomaterial are attributed to the dense DNA shell. However, the question of whether non-thiolated polyvalent DNA-AuNP could be fabricated with similar properties as its thiolated counterpart has not been addressed. This is due to a lack of fundamental understanding of DNA/AuNP interactions and the inherit challenges in applying the well-established salt-aging method to prepare such materials.

Description of the Invention

This technology describes a method for tuning the pH of the DNA/AuNP mixture, an ultrahigh capacity of non-thiolated DNA loading can be achieved in a few minutes, resulting in polyvalent DNA-AuNP conjugates with cooperative melting behaviour, a typical property for polyvalent thiolated DNA functionalized AuNPs. With this method, large AuNPs (e.g., 50 nm) can be functionalized to achieve colorimetric detection of sub-nM DNA. Further, this fast and stable DNA loading was employed to separate AuNPs of different size. This discovery not only offers a time- and cost-effective way to functionalize AuNPs with a high density of non-thiolated DNA, but also provides new insights into the fundamental understanding of how DNA strands with different sequences interact with AuNP surfaces.

Advantages

A new material that achieves similar functions for biomarkers but uses non-thiolated DNA that can reduce the cost of DNA synthesis by more than 80%. In addition, the preparation procedure is much more time-saving, taking only 5-30 min in contrast to the thiolated method, which takes about 24 hours or more to prepare the thiol DNA functionalized gold nanoparticles.

Potential Applications

Such non-thiolated DNA-functionalized gold nanoparticles will present one of the primary materials in bioanalytical chemistry, biomarkers. The reported biomarker market has grown, from $5.5 billion in 2007 to $12.4 billion in 2012 (BusinessInsights). Cancer, cardiovascular, and pathogen biomarkers are currently each worth around $1.6 billion.

Gene Expression Analysis Using Conventional and Imaging Methods

Medical imaging is the technique, process and art of creating visual representations of the interior of a body for clinical analysis and medical intervention. Medical imaging seeks to reveal internal structures hidden by the skin and bones, as well as to diagnose and treat disease. Medical imaging also establishes a database of normal anatomy and physiology to make it possible to identify abnormalities. Although imaging of removed organs and tissues can be performed for medical reasons, such procedures are usually considered part of pathology instead of medical imaging.

As a discipline and in its widest sense, it is part of biological imaging and incorporates radiology which uses the imaging technologies of X-ray radiography, magnetic resonance imaging, medical ultrasonography or ultrasound, endoscopy, elastography, tactile imaging, thermography, medical photography and nuclear medicine functional imaging techniques as positron emission tomography.

Measurement and recording techniques which are not primarily designed to produce images, such as electroencephalography (EEG), magnetoencephalography (MEG), electrocardiography (EKG), and others, but which produce data susceptible to be represented as a parameter graph *vs.* time or maps which contain information about the measurement locations, can be considered as forms of medical imaging in a limited sense.

Up until 2010, 5 billion medical imaging studies had been conducted worldwide. Radiation exposure from medical imaging in 2006 made up about 50% of total ionizing radiation exposure in the United States.

In the clinical context, "invisible light" medical imaging is generally equated to radiology or "clinical imaging" and the medical practitioner responsible for interpreting (and sometimes acquiring) the images is a radiologist. "Visible light" medical imaging involves digital video or still pictures that can be seen without special equipment.

Dermatology and wound care are two modalities that use visible light imagery. Diagnostic radiography designates the technical aspects of medical imaging and in particular the acquisition of medical images. The *radiographer* or *radiologic technologist* is usually responsible for acquiring medical images of diagnostic quality, although some radiological interventions are performed by radiologists. While radiology is an evaluation of anatomy, nuclear medicine provides functional assessment.

As a field of scientific investigation, medical imaging constitutes a sub-discipline of biomedical engineering, medical physics or medicine depending on the context: Research and development in the area of instrumentation, image acquisition (e.g. radiography), modelling and quantification are usually the preserve of biomedical engineering, medical physics, and computer science; Research into the application and interpretation of medical images is usually the preserve of radiology and the medical sub-discipline relevant to medical condition or area of medical science (neuroscience, cardiology, psychiatry, psychology, etc.) under investigation. Many of the techniques developed for medical imaging also have scientific and industrial applications.

Medical imaging is often perceived to designate the set of techniques that noninvasively produce images of the internal aspect of the body. In this restricted sense, medical imaging can be seen as the solution of mathematical inverse problems. This means that cause (the properties of living tissue) is inferred from effect (the observed signal). In the case of medical ultrasonography, the probe consists of ultrasonic pressure waves and echoes that go inside the tissue to show the internal structure. In the case of projectional radiography, the probe uses X-ray radiation, which is absorbed at different rates by different tissue types such as bone, muscle and fat.

The term noninvasive is used to denote a procedure where no instrument is introduced into a patient's body which is the case for most imaging techniques used.

Radiography

Two forms of radiographic images are in use in medical imaging; projection radiography and fluoroscopy, with the latter being useful for catheter guidance. These 2D techniques are still in wide use despite the advance of 3D tomography due to the low cost, high resolution, and depending on application, lower radiation dosages. This imaging modality utilises a wide beam of x rays for image acquisition and is the first imaging technique available in modern medicine.

- *Fluoroscopy* produces real-time images of internal structures of the body in a similar fashion to radiography, but employs a constant input of x-rays, at a lower dose rate. Contrast media, such as barium, iodine, and air are used to visualize internal organs as they work. Fluoroscopy is also used in image-guided procedures when constant feedback during a procedure is required. An image receptor is required to convert the radiation into an image after it has passed through the area of interest. Early on this was a fluorescing screen, which gave way to an Image Amplifier (IA) which was a large vacuum tube that had the receiving end coated with cesium iodide, and a mirror at the opposite end. Eventually the mirror was replaced with a TV camera.
- *Projectional radiographs*, more commonly known as x-rays, are often used to determine the type and extent of a fracture as well as for detecting pathological changes in the lungs. With the use of radio-opaque contrast media, such as barium, they can also be used to visualize the structure of the stomach and intestines - this can help diagnose ulcers or certain types of colon cancer.

Magnetic Resonance Imaging (MRI)

A magnetic resonance imaging instrument (MRI scanner), or "nuclear magnetic resonance (NMR) imaging" scanner as it was originally known, uses powerful magnets to polarise and excite hydrogen nuclei (single proton) in water molecules in human tissue, producing a detectable signal which is spatially encoded, resulting in images of the body. The MRI machine emits an RF (radio frequency) pulse that specifically binds only to hydrogen. The system sends the pulse to the area of the body to be examined. The pulse makes the protons in that area absorb the energy needed to make them spin in a different direction. This is the "resonance" part of MRI. The RF pulse makes them (only the one or two extra unmatched protons per million) spin at a specific frequency, in a specific direction.

The particular frequency of resonance is called the Larmour frequency and is calculated based on the particular tissue being imaged and the strength of the main magnetic field. MRI uses three electromagnetic fields: a very strong (on the order of units of teslas) static magnetic field to polarize the hydrogen nuclei, called the static field; a weaker time-varying (on the order of 1 kHz) field(s) for spatial encoding, called the gradient field(s); and a weak radio-frequency (RF) field for manipulation of the hydrogen nuclei to produce measurable signals, collected through an RF antenna.

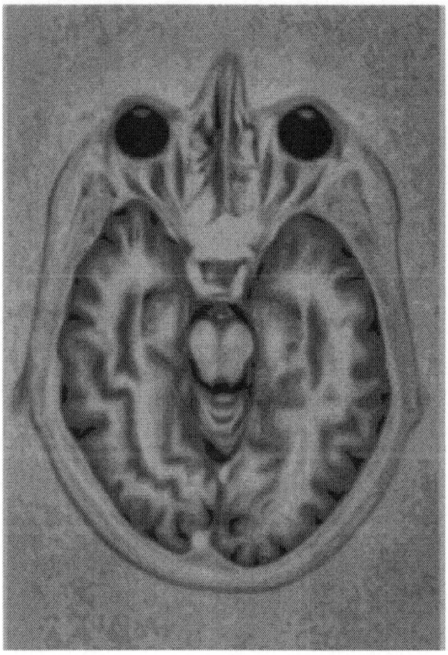

Figure: *A brain MRI representation*

Like CT, MRI traditionally creates a two dimensional image of a thin "slice" of the body and is therefore considered a tomographic imaging technique. Modern MRI instruments are capable of producing images in the form of 3D blocks, which may be considered a generalisation of the single-slice, tomographic, concept. Unlike CT, MRI does not involve the use of ionizing radiation and is therefore not associated with the same health hazards. For example, because MRI has only been in use since the early 1980s, there are no known long-term effects of exposure to strong static fields (this is the subject of some debate) and therefore there is no limit to the number of scans to which an individual can be subjected, in contrast with X-ray and CT. However, there are well-identified health risks associated with tissue heating from exposure to the RF field and the presence of implanted devices in the body, such as pace makers. These risks are strictly controlled as part of the design of the instrument and the scanning protocols used.

Because CT and MRI are sensitive to different tissue properties, the appearance of the images obtained with the two techniques differ markedly. In CT, X-rays must be blocked by some form of dense tissue to create an image, so the image quality when looking at soft tissues will be poor. In MRI, while any nucleus with a net nuclear spin can be used, the proton of the hydrogen atom remains the most widely used, especially in the clinical setting, because it is so ubiquitous and returns a large signal. This nucleus, present in water molecules, allows the excellent soft-tissue contrast achievable with MRI.

Nuclear Medicine

Nuclear medicine encompasses both diagnostic imaging and treatment of disease, and may also be referred to as molecular medicine or molecular imaging & therapeutics. Nuclear medicine uses certain properties of isotopes and the energetic particles emitted from radioactive material to diagnose or treat various pathology. Different from the typical concept of anatomic radiology, nuclear medicine enables assessment of physiology. This function-based approach to medical evaluation has useful applications in most subspecialities, notably oncology, neurology, and cardiology. *Gamma cameras* are used in e.g. scintigraphy, SPECT and PET to detect regions of biologic activity that may be associated with disease. Relatively short lived isotope, such as I^{123} is administered to the patient. Isotopes are often preferentially absorbed by biologically active tissue in the body, and can be used to identify tumors or fracture points in bone. Images are acquired after collimated photons are detected by a crystal that gives off a light signal, which is in turn amplified and converted into count data.

- *Scintigraphy* ("scint") is a form of diagnostic test wherein radioisotopes are taken internally, for example intravenously or orally. Then, gamma cameras capture and form two-dimensional images from the radiation emitted by the radiopharmaceuticals.
- *SPECT* is a 3D tomographic technique that uses gamma camera data from many projections and can be reconstructed in different planes. A dual detector head gamma camera combined with a CT scanner, which

provides localization of functional SPECT data, is termed a SPECT-CT camera, and has shown utility in advancing the field of molecular imaging. In most other medical imaging modalities, energy is passed through the body and the reaction or result is read by detectors. In SPECT imaging, the patient is injected with a radioisotope, most commonly Thallium 201TI, Technetium 99mTC, Iodine 123I, and Gallium 67Ga. The radioactive gamma rays are emitted through the body as the natural decaying process of these isotopes takes place. The emissions of the gamma rays are captured by detectors that surround the body. This essentially means that the human is now the source of the radioactivity, rather than the medical imaging devices such as X-Ray or CT.

- *Positron emission tomography* (PET) uses coincidence detection to image functional processes. Short-lived positron emitting isotope, such as ^{18}F, is incorporated with an organic substance such as glucose, creating F18-fluorodeoxyglucose, which can be used as a marker of metabolic utilisation. Images of activity distribution throughout the body can show rapidly growing tissue, like tumor, metastasis, or infection. PET images can be viewed in comparison to computed tomography scans to determine an anatomic correlate. Modern scanners may integrate PET, allowing PET-CT, or PET-MRI to optimize the image reconstruction involved with positron imaging. This is performed on the same equipment without physically moving the patient off of the gantry. The resultant hybrid of functional and anatomic imaging information is a useful tool in non-invasive diagnosis and patient management.

Fiduciary markers are used in a wide range of medical imaging applications. Images of the same subject produced with two different imaging systems may be correlated (called image registration) by placing a fiduciary marker in the area imaged by both systems. In this case, a marker which is visible in the images produced by both imaging modalities must be used. By this method, functional information from SPECT or positron emission tomography can be related to anatomical information provided by magnetic resonance imaging (MRI). Similarly, fiducial points established during MRI can be correlated with brain images generated by magnetoencephalography to localize the source of brain activity.

Ultrasound

Medical ultrasonography uses high frequency broadband sound waves in the megahertz range that are reflected by tissue to varying degrees to produce (up to 3D) images. This is commonly associated with imaging the fetus in pregnant women. Uses of ultrasound are much broader, however. Other important uses include imaging the abdominal organs, heart, breast, muscles, tendons, arteries and veins. While it may provide less anatomical detail than techniques such as CT or MRI, it has several advantages which make it ideal in numerous situations, in particular that it studies the function of moving structures in real-time, emits no ionizing radiation, and contains speckle that can be used in elastography. Ultrasound is also used as a popular research tool for capturing raw data, that can be made available through an ultrasound research interface, for the purpose of tissue characterization and implementation of new image processing techniques.

The concepts of ultrasound differ from other medical imaging modalities in the fact that it is operated by the transmission and receipt of sound waves. The high frequency sound waves are sent into the tissue and depending on the composition of the different tissues; the signal will be attenuated and returned at separate intervals. A path of reflected sound waves in a multilayered structure can be defined by an input acoustic impedance (ultrasound sound wave) and the Reflection and transmission coefficients of the relative structures. It is very safe to use and does not appear to cause any adverse effects. It is also relatively inexpensive and quick to perform. Ultrasound scanners can be taken to critically ill patients in intensive care units, avoiding the danger caused while moving the patient to the radiology department. The real time moving image obtained can be used to guide drainage and biopsy procedures. Doppler capabilities on modern scanners allow the blood flow in arteries and veins to be assessed.

Elastography

Elastography is a new imaging modality that maps the elastic properties of soft tissue. This modality emerged in the last decade. Elastography can use ultrasound, magnetic resonance imaging and tactile imaging.

Tactile Imaging

Tactile imaging is a medical imaging modality that translates the sense of touch into a digital image. The tactile image is a function of $P(x,y,z)$, where P is the pressure on soft tissue surface under applied deformation and x,y,z are coordinates where pressure P was measured. Tactile imaging closely mimics manual palpation, since the probe of the device with a pressure sensor array mounted on its face acts similar to human fingers during clinical examination, slightly deforming soft tissue by the probe and detecting resulting changes in the pressure pattern. Figure on the right presents an experiment on a composite tissue phantom examined by a tactile imaging probe illustrating the ability of tactile imaging to visualize in 3D the structure of the object.

This modality is used for imaging of the prostate, breast, vagina and pelvic floor support structures, and myofascial trigger points in muscle.

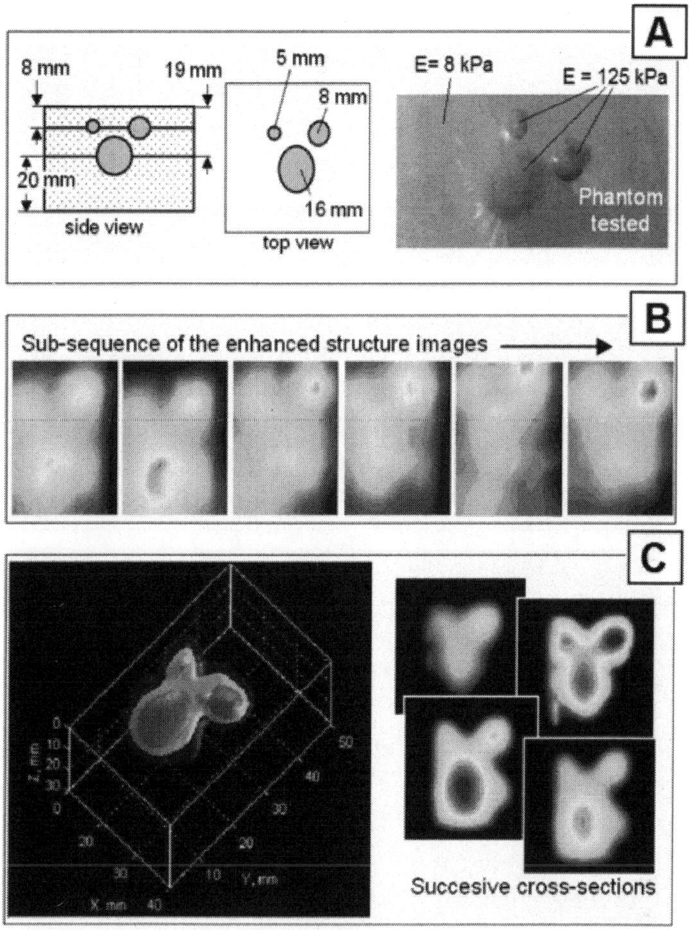

Figure: *3D tactile image (C) is composed from 2D pressure maps (B) recorded in the process of tissue phantom examination (A).*

Photoacoustic Imaging

Photoacoustic imaging is a recently developed hybrid biomedical imaging modality based on the photoacoustic effect. It combines the advantages of optical absorption contrast with ultrasonic spatial resolution for deep imaging in (optical) diffusive or quasi-diffusive regime. Recent studies have shown that photoacoustic imaging can be used in vivo for tumor angiogenesis monitoring, blood oxygenation mapping, functional brain imaging, and skin melanoma detection, etc.

Thermography

Basically used for the breast imaging. There are three approaches: the *tele-thermography*, the *contact thermography* and the *dynamic angiothermography*. These digital infrared imaging thermographic techniques are based on the principle that metabolic activity and vascular circulation in both pre-cancerous tissue and the area surrounding a developing breast cancer is almost always higher than in normal breast tissue. Cancerous tumors require an ever-increasing supply of nutrients and therefore increase circulation to their cells by holding open existing blood vessels, opening dormant vessels, and creating new ones (neo-angiogenesis theory).

Tele-thermography and contact thermography supporters claim this process results in an increase in regional surface temperatures of the breast, however there is little evidence that thermography is an accurate means of identifying breast tumours. Thermography is not approved for breast cancer screening in the United States or Canada, and medical authorities have issued warnings against thermography in both countries.

Dynamic angiothermography utilises thermal imaging but with important differences with the tele-thermography and contact thermography, that impact detection performance. First, the probes are improved over the previous liquid crystal plates; they include better spatial resolution, contrastive performance, and the image is formed more quickly. The more significant difference lies in identifying the thermal changes due to changes in vascular network to support the growth of the tumor/lesion. Instead of just recording the change in heat generated by the tumor, the image is now able to identify changes due to the vascularization of the mammary gland. It is currently used in combination with other techniques for diagnosis of breast cancer. This diagnostic method is a low cost one compared with other techniques. *The angiothermography is not a test that substitutes for other tests, but stands in relation to them as a technique that gives additional information to clarify the clinical picture and improve the quality of diagnosis.*

Tomography

Tomography is the method of imaging a single plane, or slice, of an object resulting in a tomogram. There are several forms of tomography:

- Linear tomography: This is the most basic form of tomography. The X-ray tube moved from point "A" to point "B" above the patient, while the cassette holder (or "bucky") moves simultaneously under the patient from point "B" to point "A." The fulcrum, or pivot point, is set to the area of interest. In this manner, the points above and below the focal plane are blurred out, just as the background is blurred when panning a camera during exposure. Rarely used, and has largely been replaced by computed tomography.
- Poly tomography: This was a complex form of tomography. With this technique, a number of geometrical movements were programmed, such as hypocycloidic, circular, figure 8, and elliptical. Philips Medical

Systems produced one such device called the 'Polytome.' This unit was still in use into the 1990s, as its resulting images for small or difficult physiology, such as the inner ear, was still difficult to image with CTs at that time. As the resolution of CTs got better, this procedure was taken over by the CT.

- Zonography: This is a variant of linear tomography, where a limited arc of movement is used. It is still used in some centres for visualising the kidney during an intravenous urogram (IVU).
- Orthopantomography (OPT or OPG): The only common tomographic examination in use. This makes use of a complex movement to allow the radiographic examination of the mandible, as if it were a flat bone. It is often referred to as a "Panorex", but this is incorrect, as it is a trademark of a specific company.
- Computed Tomography (CT), or Computed Axial Tomography (CAT: A CT scan, also known as a CAT scan), is a helical tomography (latest generation), which traditionally produces a 2D image of the structures in a thin section of the body. It uses X-rays. It has a greater ionizing radiation dose burden than projection radiography; repeated scans must be limited to avoid health effects. CT is based on the same principles as X-Ray projections but in this case, the patient is enclosed in a surrounding ring of detectors assigned with 500-1000 scintillation detectors (fourth-generation X-Ray CT scanner geometry). Previously in older generation scanners, the X-Ray beam was paired by a translating source and detector.
- Positron emission tomography also used in conjunction with computed tomography, PET-CT, and magnetic resonance imaging PET-MRI.

Echocardiography

When ultrasound is used to image the heart it is referred to as an echocardiogram. Echocardiography allows detailed structures of the heart, including chamber size, heart function, the valves of the heart, as well as the pericardium (the sac around the heart) to be seen. Echocardiography uses 2D, 3D, and Doppler imaging to create pictures of the heart and visualize the blood flowing through each of the four heart valves. Echocardiography is widely used in an array of patients ranging from those experiencing symptoms, such as shortness of breath or chest pain, to those undergoing cancer treatments. Transthoracic ultrasound has been proven to be safe for patients of all ages, from infants to the elderly, without risk of harmful side effects or radiation, differentiating it from other imaging modalities. Echocardiography is one of the most commonly used imaging modalities in the world due to its portability and use in a variety of applications. In emergency situations, echocardiography is quick, easily accessible, and able to be performed at the bedside, making it the modality of choice for many physicians.

Functional Near-Infrared Spectroscopy

Functional Near-Infrared Spectroscopy (fNIR or fNIRS), is the use of NIRS (near-infrared spectroscopy) for the purpose of functional neuroimaging. Using fNIR, brain activity is measured through hemodynamic responses associated with neuron behaviour.

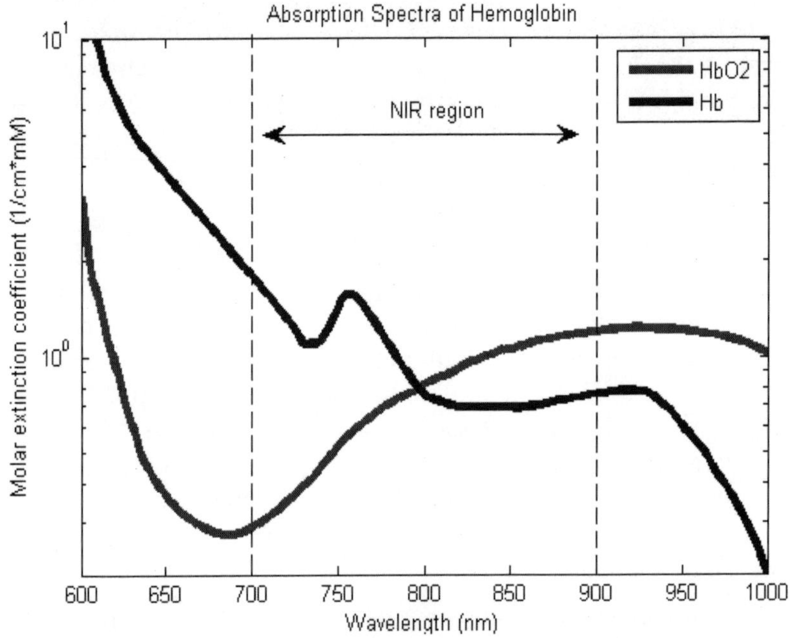

Figure: *Absorption spectra for oxy-Hb and deoxy-Hb for Near-infrared wavelengths*

fNIR is a non-invasive imaging method involving the quantification of chromophore concentration resolved from the measurement of near infrared (NIR) light attenuation, temporal or phasic changes. NIR spectrum light takes advantage of the optical window in which skin, tissue, and bone are mostly transparent to NIR light in the spectrum of 700-900 nm, while hemoglobin (Hb) and deoxygenated-hemoglobin (deoxy-Hb) are stronger absorbers of light. Differences in the absorption spectra of deoxy-Hb and oxy-Hb allow the measurement of relative changes in hemoglobin concentration through the use of light attenuation at multiple wavelengths.

Two or more wavelengths are selected, with one wavelength above and one below the isosbestic point of 810 nm at which deoxy-Hb and oxy-Hb have identical absorption coefficients. Using the modified Beer-Lambert law (mBLL), relative concentration can be calculated as a function of total photon path length. Typically the light emitter and detector are placed ipsilaterally on the subjects skull so recorded measurements are due to back-scattered (reflected) light following elliptical pathways.

The use of fNIR as a functional imaging method relies on the principle of neuro-vascular coupling also known as the Haemodynamic response or BOLD (Blood-Oxygenation-Level-Dependent) response. This principle also forms the core of fMRI techniques.

Through neuro-vascular coupling, neuronal activity is linked to related changes in localized cerebral blood flow. fNIR and fMRI are sensitive to similar physiologic changes and are often comparative methods. Studies relating fMRI and fNIR show highly correlated results in cognitive tasks. fNIR has several advantages in cost and portability over fMRI, but cannot be used to measure cortical activity more than 4 cm deep due to limitations in light emitter power and has more limited spatial resolution. fNIR includes the use of Diffuse Optical Tomography (DOT/NIRDOT) for functional purposes.

Spectroscopic Techniques

There are four current methods of fNIR Spectroscopy. Multiplexing fNIR channels can allow 2D topographic functional maps of brain activity (ex: Hitatchi ETG-4000) while using multiple emitter spacings may be used to build 3D tomographic maps.

Frequency Domain (FD)

In frequency domain systems, NIR laser sources provide an amplitude modulated sinusoid at frequencies of several megahertz. Changes in the back-scattered signal's amplitude and phase provide information on changes in the concentration of hemoglobin and provide direct measurement of absorption coefficients obviating the need for information about photon path-length.

Because of the need for modulated lasers as well as phasic measurements, frequency domain systems are more technically complex, less portable, and more expensive than continuous wave systems. However these systems are capable of providing absolute concentrations of Hb and deoxy-Hb.

Time-Resolved (TR)

In time-resolved spectroscopy, a very short NIR pulse is introduced with a pulse length usually on the order of picoseconds. Through time-of-flight measurements, photon path-length may be directly observed by dividing resolved time by the speed of light. Because of the need for high-speed detection and high-speed emitters, time-resolved methods are the most expensive and technically complicated method. Information about hemodynamic changes can be found in the attenuation, decay, and time profile of the back-scattered signal.

Spatially-Resolved Spectroscopy (SRS)

SRS systems use localized gradients in light attenuation to determine absolute ratios of oxy-Hb and deoxy-Hb. Using a spatial measurement, SRS systems do not require knowledge of photon path-length to make this calculation, however measured concentrations of oxy-Hb and deoxy-Hb are relative to the unknown coefficient of scattering in the media. This technique is most commonly used in cerebral oxymetry systems that report a Tissue Oxygenation Index (TOI).

Applications

Brain Computer Interface (BCI)

fNIR has been successfully implemented as a control signal for BCI systems.

Brain Imaging

fNIR has been widely accepted as a brain imaging technique. Researchers at the University of Pittsburgh Schools of the Health Sciences in Pittsburgh, Pennsylvania, led by Dr. Anthony Kontos, Ph.D., Assistant Research Director, UPMC Sports Medicine Concussion Program, and Assistant Professor, Pitt Department of Orthopedic Surgery, have conducted a National Institutes of Health-funded preliminary study (nine people between ages 18 to 45 who were symptomatic within 15 days to 1 1/2 months of a sports-related concussion), and have done large-scale statistical analysis, using portable fNIR, for use as a relatively low-cost evaluative tool in concussions.

Image Gently and Image Wisely Campaigns

In response to increased concern by the public over radiation doses and the ongoing progress of best practices, The Alliance for Radiation Safety in Pediatric Imaging was formed within the Society for Pediatric Radiology. In concert with The American Society of Radiologic Technologists, The American College of Radiology and The American Association of Physicists in Medicine, the Society for Pediatric Radiology developed and launched the Image Gently Campaign which is designed to maintain high quality imaging studies while using the lowest doses and best radiation safety practices available on pediatric patients. This initiative has been endorsed and applied by a growing list of various Professional Medical organisations around the world and has received support and assistance from companies that manufacture equipment used in Radiology.

Following upon the success of the Image Gently campaign, the American College of Radiology, the Radiological Society of North America, the American Association of Physicists in Medicine and the American Society of Radiologic Technologists have launched a similar campaign to address this issue in the adult population called Image Wisely. The World Health Organisation and International Atomic Energy Agency (IAEA) of the United Nations have also been working in this area and have ongoing projects designed to broaden best practices and lower patient radiation dose.

Maximizing Imaging Procedure Use

The amount of data obtained in a single MR or CT scan is very extensive. Some of the data that radiologists discard could save patients time and money, while reducing their exposure to radiation and risk of complications from invasive procedures.

Creation of Three-Dimensional Images

Recently, techniques have been developed to enable CT, MRI and ultrasound scanning software to produce 3D images for the physician. Traditionally CT and MRI scans produced 2D static output on film. To produce 3D images, many scans are made, then combined by computers to produce a 3D model, which can then be manipulated by the physician. 3D ultrasounds are produced using a somewhat similar technique. In diagnosing disease of the viscera of abdomen, ultrasound is particularly sensitive on imaging of biliary tract, urinary tract and female reproductive organs (ovary, fallopian tubes). As for example, diagnosis of gall stone by dilatation of common bile duct and stone in common bile duct. With the ability to visualize important structures in great detail, 3D visualization methods are a valuable resource for the diagnosis and surgical treatment of many pathologies. It was a key resource for the famous, but ultimately unsuccessful attempt by Singaporean surgeons to separate Iranian twins Ladan and Laleh Bijani in 2003. The 3D equipment was used previously for similar operations with great success.

Diffuse Optical Tomography

Diffuse optical imaging (DOI) is a method of imaging using Near Infrared Spectroscopic (NIRS) or Fluorescence based methods. The technique has many applications to neuroscience, sports medicine, wound monitoring, and cancer detection. Typically DOI techniques monitor changes in concentrations of oxygenated and deoxygenated hemoglobin and may additionally measure redox states of cytochromes. The technique may also be referred to as diffuse optical tomography (DOT), near infrared optical tomography (NIROT) or fluorescence diffuse optical tomography (FDOT), depending on the usage.

When used to create 3D volumetric models of the imaged material DOI is referred to as diffuse optical tomography, whereas 2D imaging methods are classified as diffuse optical topography.

In neuroscience, functional measurements made using NIR wavelengths, DOI techniques may classify as functional near infrared spectroscopy fNIRS.

Diffusive Optical Imaging in Neuroscience

Diffusive Optical Imaging (DOI) also known as *Near-Infrared Optical tomography* (NIROT), *Diffuse Optical Tomography* (DOT) or *Optical Diffusion Tomography* (ODT) is a technique that gives neuroscientists the ability to simultaneously obtain information about the source of neural activity as well as its time course, allowing to "see" neural activity and study the functioning of the brain.

In this method, a near-infrared laser is positioned on the scalp. Detectors composed of optical fiber bundles are located a few centimetres away from the light source. These detectors sense how the path of light is altered, either through absorption or scattering, as it traverses brain tissue.

This method can provide two types of information. First, it can be used to measure the absorption of light, which is related to concentration of chemicals in the brain. Second, it can measure the scattering of light, which is related to physiological characteristics such as the swelling of glia and neurons that are associated with neuronal firing.

Typical applications include rapid 2D optical *topographic* imaging of the event-related optical signal (EROS) or Near-infrared spectroscopy (NIRS) signal following brain activity and *tomographic* reconstruction of an entire 3D volume of tissue to diagnose breast cancer or neonatal brain haemorrhage. The spatial resolution of Diffuse Optical Tomography (DOT) techniques is several millimetres, comparable to the lower end of functional magnetic resonance imaging (fMRI). The temporal resolution of EROS is very good, comparable to electroencephalography, and magnetoencephalography (~milliseconds), while that of NIRS, which measures hemodynamic changes rather than neuronal activity, is comparable to fMRI (~seconds). DOT instruments are relatively low cost ($150,000), portable and immune to electrical interference. The signal-to-noise ratio of NIRS is quite good, enabling detection of responses to single events in many cases. EROS signals are much weaker, typically requiring averaging of many responses.

Important chemicals that this method can detect include hemoglobin and cytochromes.

Gene Expression

Gene expression is the process by which information from a gene is used in the synthesis of a functional gene product. These products are often proteins, but in non-protein coding genes such as ribosomal RNA (rRNA), transfer RNA (tRNA) or small nuclear RNA (snRNA) genes, the product is a functional RNA. The process of gene expression is used by all known life - eukaryotes (including multicellular organisms), prokaryotes (bacteria and archaea), and utilised by viruses - to generate the macromolecular machinery for life. Several steps in the gene expression process may be modulated, including the transcription, RNA splicing, translation, and post-translational modification of a protein. Gene regulation gives the cell control over structure and function, and is the basis for cellular differentiation, morphogenesis and the versatility and adaptability of any organism. Gene regulation may also serve as a substrate for evolutionary change, since control of the timing, location, and amount of gene expression can have a profound effect on the functions (actions) of the gene in a cell or in a multicellular organism.

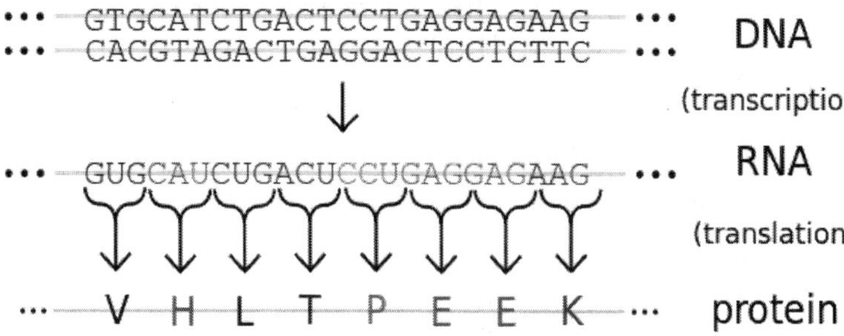

Figure: *Genes are expressed by being transcribed into RNA, and this transcript may then be translated into protein.*

In genetics, gene expression is the most fundamental level at which the genotype gives rise to the phenotype. The genetic code stored in DNA is "interpreted" by gene expression, and the properties of the expression give rise to the organism's phenotype. Such phenotypes are often expressed by the synthesis of proteins that control the organism's shape, or that act as enzymes catalysing specific metabolic pathways characterising the organism.

Imaging Gene Expression in Live Cells and Tissues

Monitoring gene expression is crucial for studying the responses of gene therapy and clarifying gene function in various environments. Molecular imaging is a powerful tool for noninvasive visualization of gene expression. This article summarizes the current status of fluorescence and bioluminescence imaging (BLI) of gene expression in live cells and tissues, with the emphasis mainly on the early studies that pioneered the field. First, we describe fluorescence imaging of gene expression with a wide variety of fluorescent proteins. Next, we discuss the strategies for BLI of gene expression.

Besides incorporating the reporter gene into the host DNA, mRNA-based BLI of gene expression is also briefly mentioned. Last, the construction of double- and triple-fusion reporter genes is presented. Because no single imaging modality is perfect and sufficient to obtain all of the necessary information for a given question, combinations of multiple molecular imaging modalities can offer synergistic advantages over any modality alone. Noninvasive optical imaging of gene expression has revolutionized biomedical research, and the progress made over the last decade should allow molecular imaging to play a major role in the field of gene therapy. For basic and preclinical research, optical imaging is indispensable for imaging gene expression. However, for clinical imaging of gene expression, positron emission tomography (PET) holds the greatest promise.

Overview

With the ability to engineer genes and to create knockin and knockout models of human diseases, it has become clear that gene therapy can play an important role in disease management. Over the last four decades, gene therapy has moved from preclinical to clinical studies for diseases ranging from monogenic recessive disorders (e.g., hemophilia) to more complex diseases, such as cancer, cardiovascular disorders, and human immunodeficiency virus (HIV) infection. To choose the appropriate gene therapy strategy and to optimize the therapeutic efficacy, an effective monitoring system is needed. One of the most economical and practical approaches is to adopt certain molecular imaging techniques for monitoring gene expression in vivo.

In recent years, noninvasive molecular imaging has emerged as a powerful tool for monitoring cellular and molecular events in vivo. A variety of imaging technologies is being investigated as tools for evaluating gene therapy efficiency and for studying gene expression in living subjects. Noninvasive, longitudinal, and quantitative imaging of gene expression can help in human gene therapy trials and can also facilitate preclinical experimental studies in animal models. Radionuclide-based imaging techniques (i.e., single-photon emission computed tomography [SPECT] and PET) have the greatest clinical potential and offer many advantages for noninvasive imaging of gene expression.

Besides imaging gene vectors to indirectly visualize the gene expression efficiency, PET and SPECT can play a significant role in imaging gene expression using diverse reporter genes and reporter probes (Kang and Chung 2008). If transcription of a reporter gene is induced, translation of the reporter gene mRNA will lead to a protein product that can interact with the imaging reporter probe (administered in trace amounts for PET/SPECT applications; hence the term "tracer"). This interaction may be based on the intracellular enzymatic conversion of the reporter probe with the retention of the metabolite(s), or a receptor–ligand-based interaction.

Examples of intracellular reporters include the herpes simplex virus type 1 thymidine kinase (HSV1-tk) and its mutant gene (HSV1-sr39tk). Substrates that have been studied to date, as PET reporter probes for HSV1-tk, can be classified into two main categories: pyrimidine nucleoside derivatives and acycloguanosine derivatives. A few examples of reporters on the surface of cells include the dopamine 2 receptor (D2R), receptors for the human type 2 somatostatin receptor (hSSTr2), and the sodium iodide symporter (NIS).

Although magnetic resonance imaging (MRI) has relatively low sensitivity, it can be used to image gene expression in vivo. MRI has been used for imaging iron-related gene products (e.g., transferrin), certain enzymes (e.g., â-galactosidase [â-gal]), and chemical exchange saturation transfer-related products (e.g., polylysine-containing proteins).

Whereas PET/SPECT and MRI offer greater clinical potential than many other imaging modalities, optical imaging techniques, such as fluorescence and bioluminescence, have the following advantages: They are more economical, are easier to handle, do not need any radioisotopes, and are quite sensitive in certain scenarios. Therefore, optical imaging has been widely used for monitoring gene expression in live cells and tissues. In the remainder of this article, we summarize the current state of the art in optical imaging of gene expression.

Fluorescence Imaging of Gene Expression

For fluorescence imaging of gene expression, green fluorescent protein (GFP; emission peak at 509 nm) and its variants are the mainstay among fluorescent probes. Because light at longer wavelengths (i.e., red or near infrared) penetrates tissues better than light at shorter wavelengths, much effort has been devoted to engineering fluorescent proteins that can emit in the red or near infrared.

Fluorescence imaging of gene expression. (*A*) The structure of GFP. (*B*) A transgenic GFP mouse. (*C*) Real-time whole-body imaging of a GFP-expressing human glioma growing in the brain of a nude mouse at 1 wk (*left*), 3 wk (*centre*), and 5 wk (*right*) after surgical orthotopic implantation. (*D*) Monomeric and tandem dimeric fluorescent proteins derived from *Aequorea* GFP or *Discosoma* RFP, expressed in bacteria, and purified.

GFP-Based Imaging

From the early 1990s, GFP and its variants have become invaluable markers for monitoring protein localization and gene expression in vivo. One early example of imaging gene expression with GFP was the demonstration of vascular endothelial growth factor (VEGF) promoter activity. The VEGF promoter was linked to the GFP gene, and the expression of GFP was imaged by intravital microscopy. For such intravital studies, in which only limited tissue penetration of the reporter protein is needed, GFP was found to be readily detectable, confined to the cell, and stable (with a half-life of up to 24 h in cells). Although GFP penetrates tissue poorly, its expression can be visualized noninvasively in intact animals. GFP has also been used frequently in stem cell research to monitor gene expression. Modifications of GFP to increase its signal intensity and thermostability, as well as to alter its emission spectrum, have expanded the utility of GFP in gene transfer studies.

GFP is very useful for imaging tumor-related gene expression. Mouse models of metastatic cancer were developed with genetically fluorescent tumors, in which the GFP gene was cloned into cancer cell lines and selected for stable GFP expression, which was imaged in fresh tissue, both in situ and externally. With this model system, tumor location and metastasis were detected and were visualized in situ within host organs down to the single-cell level. Imaging GFP-transfected tumor cells was a fundamental advance in visualizing tumor growth and metastasis in real time in vivo. In addition, tumor growth and metastatic development and inhibition of these processes by certain drugs could now be imaged and quantified for rapid screening of anticancer drugs. Research about imaging tumor-related gene expression has flourished, including the use of a reporter gene system consisting of enhanced green fluorescent protein (eGFP) and wild-type HSV1-tk for both fluorescence and PET imaging.

GFP can also serve as a powerful tool for imaging other diseases, such as HIV infection. For example, the function of the HIV-1 Tat protein (capable of traversing cell membranes) has been elucidated by fluorescence imaging. After exogenously adding Tat-GFP fusion protein to live HeLa and CHO cells, it was found that the internalization process of full-length Tat, as well as of heterologous proteins fused to the transduction domain of Tat, exploits a caveolar-mediated pathway. With fluorescence imaging, the dynamic movement of individual GFP-tagged, Tat-filled caveolae toward the nucleus was observed directly.

Red Fluorescent Protein-Based Imaging

In 1999, a new red fluorescent protein (RFP) was isolated from tropical corals and was termed DsRed. With emission maxima at 509 and 583 nm, respectively, eGFP and DsRed are well suited for virtually crossover-free, dual-colour imaging upon simultaneous excitation. Mixed populations of *Escherichia coli* expressing either eGFP or DsRed were imaged by one-photon and two-photon microscopy. Both excitation modes were found to be suitable for imaging cells expressing either of the fluorescent proteins.

The predominant use of DsRed has been for multicolour imaging in plants, together with the GFP variants, because its red-shifted excitation/emission spectra avoid damaging plant cells and tissues by the excitation light. Dual gene expression has also been imaged in animal models with GFP and DsRed-2 after transduction with a dual-promoter lentiviral vector. RFP-based gene expression imaging has also been used extensively in stem cell research. For example, the whole-body biodistribution and persistence of multipotent adult progenitor cells, transfected with DsRed-2, were visualized in vivo and in tissue sections. Another important field for RFP-based research is cancer cell imaging. For example, fluorescent pancreatic cancer cells expressing a high level of the DsRed-2 gene were used to establish an orthotopic metastatic pancreatic cancer model. The high-level expression of DsRed-2 enabled noninvasive imaging of distant micrometastases in their target organs, even in deep tissue such as the lung.

Wild-type DsRed has several drawbacks including inefficient folding of the protein, extremely slow maturation of the chromophore, and tetramerization even in dilute solutions. Therefore, stepwise evolution of DsRed to a dimer and then to a true monomer (designated as mRFP1) was performed. The monomer and DsRed show similar brightness in living cells. In addition, the excitation and emission peaks of mRFP1, 584 and 607 nm, respectively, are ~25-nm red-shifted from DsRed, conferring greater tissue penetration and spectral separation from autofluorescence and other fluorescent proteins. Subsequently, a series of monomeric fluorescent proteins has been reported, which further expanded the toolbox for fluorescence imaging of gene expression.

Infrared Fluorescent Protein (IFP)

Generally speaking, GFP remains the first choice for fluorescence imaging of gene expression because it is relatively easy to fuse it with other genes and it has enhanced fluorescence characteristics compared with RFP. However, for optimal performance in vivo imaging, near-infrared fluorescent (NIRF; 700–900 nm) proteins are superior. The NIRF signal can pass through deeper tissues, and the background fluorescence in this part of the spectrum is very low.

Recently, it was reported that a bacteriophytochrome from *Deinococcus radiodurans*, incorporating biliverdin as the chromophore, could be engineered into a monomeric IFP with excitation and emission maxima of 684 and 708 nm, respectively. The IFPs express well in mammalian cells and mice and spontaneously incorporate biliverdin, which is ubiquitous as the initial intermediate in heme catabolism but has negligible fluorescence by itself. These IFPs provided the basis for further engineering of IFPs with better performance characteristics, which is expected to find broad use in future biomedical research, in particular, those related to gene expression imaging.

β-Galactosidase-Based Imaging

Besides the fusion of fluorescent proteins with target gene products, several other strategies can also be adopted for fluorescence imaging of gene expression. The use of fluorogenic substrates for certain gene translation products (e.g., enzymes and proteins) is one choice. For example, the bacterial *lacZ* gene, which encodes β-gal, is a common reporter gene used in transgenic mice. However, the absence of fluorogenic substrates usable in live animals greatly hampered the applications of this reporter gene. In 2007, a far-red fluorescent substrate, 9H-(1,3-dichloro-9,9-dimethylacridin-2-one-7-yl) â-D-galactopyranoside, was developed for imaging β-gal expression. With â-gal as a reporter of tumor growth, as few as 1000 â-gal-expressing tumor cells located under the skin could be detected with this substrate.

Tomographic Imaging of Gene Expression

Aside from the various strategies for imaging gene expression in live cells and tissues by fluorescence, as mentioned above, there have also been significant advances in imaging instrumentation. To generate a more quantitative fluorescence readout, fluorescence diffuse tomography has been developed recently for small animal imaging. The animal is scanned in the transilluminative configuration by a single source and a detector pair, and a reconstruction algorithm was developed to estimate the fluorophore distribution. Another technique, fluorescence-mediated tomography (FMT), can also provide important insights into in vivo gene expression within deep tissues.

With the wide variety of tools for fluorescence imaging of gene expression, multiplexing is highly desirable and certainly possible. Using an imaging system capable of spectral unmixing, in which fluorophores with different emission spectra can be readily separated and analyzed simultaneously within the same sample, multiple genes fused to different fluorescent proteins can be imaged at the same time, shedding new light into the biology and the mechanisms of various diseases. With newly developed fluorescent proteins that emit at longer wavelengths and the constantly evolving imaging systems, which can be used clinically to image certain sites of the human body (e.g., tissues close to the skin, accessible by endoscopy, and/or during surgery), fluorescence imaging of gene expression serves as an invaluable research tool for cell- and animal-based studies, as well as for certain potential clinical applications.

Bioluminescence Imaging of Gene Expression

Another optical imaging technique, BLI, is an attractive alternative to fluorescence for optical imaging of gene expression. Although BLI has little, if any, clinical potential, it is virtually free of any background signal, an ideal property for sensitive imaging of gene expression in cells and small animals. Bioluminescence is the emission of light from biochemical reactions that occur within a living organism. The luciferases are a family of photoproteins that can be isolated from a variety of insects, marine organisms, and prokaryotes. Luciferase has been used as a reporter gene in transgenic mice, but, until the instrumentation was created for BLI, the detection of luciferase activity required either sectioning the animal or excising the tissue and homogenizing it to measure the luciferase activity in a luminometre. BLI has proven to be a very powerful method for detecting luciferase activity in intact animal models. It is noninvasive, convenient, and relatively inexpensive, thus making it an excellent method for elucidating the pathobiology of diseases such as inflammation/injury, infection, and cancer in animal models. To date, firefly luciferase (Fluc) and *Renilla* luciferase (Rluc) are the most widely used reporter proteins for BLI of gene expression in living animals. Another type of luciferase, the bacterial luciferase, is primarily used in the study of infection because it is limited to bacteria that express the reporter gene and produce the substrate.

Bioluminescence imaging of gene expression. (*A*) The bioluminescent reaction catalyzed by Fluc. (*B*) The bioluminescent reaction catalyzed by Rluc. (*C*) The magnitude and the duration of Fluc gene expression in different host immune systems can be monitored noninvasively and repetitively using a cooled CCD camera. For Swiss Webster mice, the Fluc activity was highest at day 2 and significantly dropped over time. For nude mice, a considerable amount of Fluc activity was seen throughout the study period.

Fluc-Based BLI

Light emission (yellow–green colour, 557 nm) from the firefly *Photinus pyralis*, which is generally believed to be the most efficient bioluminescence system known to date, makes Fluc an excellent tool for monitoring gene expression. Fluc has been used to image gene expression in cells since the early 1990s. In 2001, a method was described for repetitively tracking in vivo gene expression of Fluc in skeletal muscles of mice with a cooled charge-coupled device (CCD) camera. The in vivo bioluminescence signals correlated well with results from in vitro luciferase enzyme assays, which showed the ability of BLI to sensitively and noninvasively track the location, the magnitude, and the persistence of Fluc-related gene expression.

Fluc has broad applications in monitoring disease-related gene expression. For example, the *cyclooxygenase-2* (*COX-2*) gene plays a role in a wide variety of physiologic pathways and is a major target for therapeutic intervention in many pathophysiologic contexts such as pain, fever, inflammation, and cancer. Expression of the *COX-2* gene can be induced in a wide range of cells, in response to an ever-increasing number of stimuli. BLI was successfully performed to image the expression of the Fluc gene in tumor xenografts, which were stably transfected with a chimeric gene containing the first kilobase of the murine *COX-2* promoter. The imaging data suggested that gene expression from the *COX-2* promoter can be easily analyzed in a variety of disease models in which the *COX-2* gene is up-regulated. BLI of gene expression with Fluc was also tested in larger animals such as rabbits (Li et al. 2005). It was reported that the BLI signal was capable of passing through at least 1 cm of muscle tissue.

BLI with Fluc is also applicable for the evaluation of DNA vaccines. Administration of naked DNA into animals has been used as a research tool for developing DNA vaccines. To monitor the distribution and the duration of gene expression of a DNA vaccine in living subjects, the naked DNA encoding Fluc was used as an imaging reporter gene in a mouse model. It was concluded that BLI with Fluc could be useful for monitoring the location, the intensity, and the duration of gene expression of naked DNA vaccines in living animals, both noninvasively and repetitively. Similar to the study of DNA vaccines, BLI with Fluc (which serves as a model gene) can also be used to evaluate the gene delivery efficiency of certain gene delivery systems, especially in tumor models.

One major limitation of Fluc is that the light generated is in the yellow–green range, which has poor tissue penetration. In 2005, a set of red- and green-emitting luciferase mutants was reported. The bioluminescence properties of these mutants are suitable for expanding the use of the *P. pyralis* system in dual-colour reporter assays, biosensor measurements with internal controls, and imaging. Using a combination of mutagenesis methods, a red-emitting luciferase with a bioluminescence maximum of 615 nm was created, which also has a narrow emission bandwidth and favourable kinetic properties. Studies in animal models showed that these luciferases could detect gene expression at the attomole level, many orders of magnitude more sensitive than Fluc.

Rluc-Based BLI

Rluc is another promising bioluminescence reporter. Distinct from Fluc in terms of its origin, enzyme structure, and substrate requirements (Inouye and Shimomura 1997), Rluc was isolated from the sea pansy (*Renilla reniformis*), which displays blue–green bioluminescence upon stimulation. It can catalyze the oxidation of coelenterazine, which leads to a bioluminescence signal (Fig. 2B). Rluc was cloned and sequenced in 1991 (Lorenz et al. 1991) and has been used as a marker for gene expression in bacteria, yeast, plant, and mammalian cells (Lorenz et al. 1996).

BLI of gene expression with Rluc dates back to the early 2000s. In one pioneering report, Rluc was used for noninvasive BLI of advanced human prostate cancer lesions in living mice by a targeted gene transfer vector (Adams et al. 2002). In another early study, the ability to image gene expression based on Rluc bioluminescence was validated by injecting the substrate coelenterazine into living mice (Bhaumik and Gambhir 2002). Cells transiently expressing the Rluc gene, located in the peritoneum, subcutaneous layer, as well as in the liver and the lungs of living mice, were imaged after tail-vein injection of coelenterazine. Importantly, both Rluc and Fluc expression can be imaged in the same living mouse, although the kinetics of light production are distinct for the two enzymes. The imaging strategy validated in this study has direct application in various studies in which two molecular events need to be tracked, such as the trafficking of two cell populations, two gene therapy vectors, or indirect monitoring of two endogenous genes.

In a follow-up study, the expression in live mice of a novel synthetic Rluc reporter gene (hRluc) was explored; it has previously been reported to be a more sensitive reporter in mammalian cells than the native Rluc. It was found that hRluc:coelenterazine yielded stronger signals than the Fluc:D-luciferin in both cell culture and live animal studies.

Regarding the prognosis of cancer treatment, early detection of the tumor and its metastases is crucial. In animal models, Rluc can be used for detecting the expression of diagnostic genes in different stages of tumor development, thus delineating the location of primary and metastatic tumors. To achieve noninvasive measurement of chemotherapy-induced changes in the expression of genes related to tumor growth, BLI was used to image the alteration in human telomerase reverse transcriptase (hTERT) gene expression in tumor cells before and after 5-fluorouracil treatment. Several fusion reporters directed by the hTERT promoter fragments were investigated, which integrate the hRluc (for BLI), the mRFP1 (for fluorescence imaging), and a truncated thymidine kinase (for PET imaging with radiolabelled acycloguanosines), respectively. In vitro studies showed that although all three of the reporter systems can visualize the hTERT promoter activity, BLI was the most powerful one, and it could also provide assistance in choosing the optimal tracers for PET imaging.

BLI has been widely adopted for stem-cell-based studies. The discovery of human embryonic stem cells (hESCs) has dramatically expanded the tools available to scientists and clinicians in the field of regenerative medicine. However, direct injection of hESCs, and cells differentiated from hESCs, into living organisms was hampered by significant cell death, teratoma formation, and host immune rejection. Understanding the hESC behaviour in vivo after transplantation requires noninvasive imaging techniques, such as BLI, to longitudinally monitor the hESC localization, proliferation, and viability. Bioluminescence reporter genes can be transcribed either constitutively or under specific biological or cellular conditions, depending on the type of promoter used. Stably transduced cells that carry the reporter construct within their chromosomal DNA will pass the reporter construct DNA to daughter cells, allowing for longitudinal monitoring of hESC survival and proliferation in vivo with BLI. Because expression of the reporter gene product is required for signal generation, only viable parent and daughter cells will generate a BLI signal, whereas the apoptotic or dead cells cannot, which is the key advantage of reporter gene-based cell tracking over direct cell labelling.

BLI of mRNA

In addition to incorporating Fluc or Rluc into DNA, a strategy for imaging mRNA with BLI was designed using spliceosome-mediated RNA *trans*-splicing (SMaRT). SMaRT provides an effective means for reprogramming mRNAs and the proteins they encode. It can have a broad range of applications, including RNA repair and molecular imaging, each governed by the nature of the sequences delivered by the pre-*trans*-splicing molecule (PTM). In one groundbreaking study, the ability of SMaRT to optically image the expression of an exogenous gene at the level of pre-mRNA splicing was shown in both cells and living animals. Because of the modular design of PTMs, there is potential for SMaRT to be used for imaging the expression of any arbitrary gene of interest in living subjects.

Recently, this strategy was improved by using signal amplification and a facile method of delivery for developing a class of generalized probes capable of imaging pre-mRNA in a sequence-specific manner. Incorporating a modular binding domain that confers specificity by base-pair complementarity to the target pre-mRNA, the PTMs were designed to target a chimeric target minigene and *trans*-splice the Rluc gene onto the end of the target. After hydrodynamic delivery of the PTMs and target genes in mice, the efficiency and the specificity of the *trans*-splicing reaction were found to vary depending on the binding domain length and the structure. Nonetheless, specific *trans*-splicing was observed in living animals that showed a proof of principle for a generalized imaging probe against RNA, which can amplify the signal upon detection and can be delivered with existing gene delivery methodology.

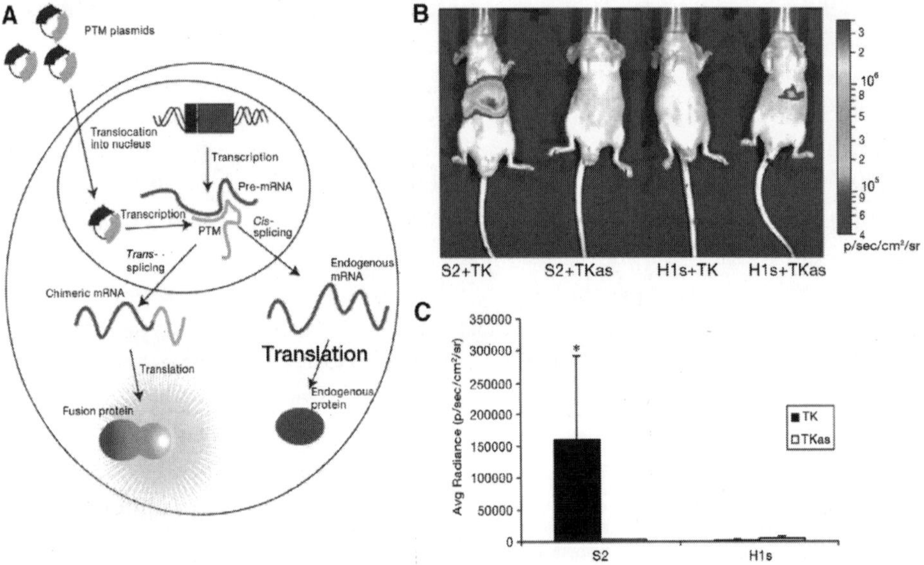

BLI of mRNA. (*A*) A schematic of the SMaRT imaging strategy. (*B*) Representative images of nude mice injected hydrodynamically with a combination of PTM and target plasmids, S2 + TK:PTM, and the target gene. The remaining three mice are negative controls. (*C*) Average radiance for each mouse (mean ± SEM) in photons/sec per cm²/steradian.

BLI of gene expression has been used in a wide variety of disease models over the last decade, and the above-mentioned studies serve only as representative examples. Theoretically, Fluc or Rluc can be fused to almost any gene of interest, and the expression of the gene can be noninvasively monitored with BLI. One major advantage of BLI is that it is quite quantitative.

Therefore, the changes in the bioluminescence signal can reflect the differences in gene expression level, provided that all other related variables remain constant (e.g., the depth of the tissue). Over the years, BLI has been used in a variety of model systems for applications ranging from stem cell trafficking to cancer therapy to drug screening. Bear in mind, however, that BLI is exclusively a research tool in preclinical studies and cannot be used in humans.

Multimodality Imaging of Gene Expression

Among all of the molecular imaging modalities, no single modality is sufficient to obtain all of the necessary information for a given question. For example, it is difficult to accurately quantify fluorescence signals in living subjects with fluorescence imaging alone, particularly in deep tissues; BLI is quantitative but is not suitable for clinical studies; and radionuclide-based imaging techniques, such as PET, are very sensitive but have relatively poor spatial resolution. A combination of multiple molecular imaging modalities can offer synergistic advantages over any modality alone.

Various double-fusion and triple-fusion reporter genes have been reported for multimodality imaging of gene expression. In one early study, a reporter vector encoding a mutant HSV1-sr39tk and Rluc was constructed (Ray et al. 2003). The two genes were joined by a 20-amino-acid spacer sequence. Both PET and BLI were able to delineate tumors stably expressing the fusion gene in live xenograft-bearing mice.

Subsequently, several triple-fusion reporter genes that can monitor bioluminescence, fluorescence, and PET imaging have been constructed. A triple-fusion reporter vector consisting of Rluc, RFP, and HSV1-sr39tk (ttk) was found to confer activity of every protein composition in cell culture (Ray et al. 2004). In xenograft mouse models, the lentiviral vector encoding the triple-fusion reporter gene also showed a good correlation of signals from different imaging modalities. Furthermore, the vector expression lasted for 40–50 d in live mice.

To exploit the combined strengths of each imaging technique and to facilitate multimodality imaging, a dual-reporter construct was established in which Fluc was fused in-frame to the N terminus of a mutant HSV-tk kinetically enhanced for PET (Kesarwala et al. 2006). In addition, a triple-reporter construct was developed in which monster GFP was introduced into the fusion vector downstream from an internal ribosome entry site to allow analysis by fluorescence microscopy or by flow cytometry without compromising the specific activities of the upstream fusion components. In mice, somatic gene transfer of a ubiquitin promoter-driven triple-fusion plasmid showed a >1000-fold increase in liver photon flux and a more than twofold increase in liver retention of ^{18}F-FHBG by micro-PET when compared with mice treated with a control plasmid.

Construction and validation of several triple-fusion genes—each composed of a bioluminescent, a fluorescent, and a PET reporter gene in cell culture and in living subjects—were further explored (Ray et al. 2007). A mutant of a thermostable Fluc, bearing the peroxisome localization signal, was designed to have greater cytoplasmic localization and improved access for its substrate, D-luciferin. This mutant showed severalfold higher activity than Fluc both in vitro and in vivo. The improved version of the triple-fusion vector showed a significantly higher bioluminescence signal than the previous triple-fusion vectors. As the third reporter component of this triple-fusion vector for PET applications, a truncated version of wild-type HSV1-tk also retained a higher expression level than the truncated mutant HSV1-sr39tk. It was suggested that this improved triple-fusion reporter vector should enable high-sensitivity detection of lower numbers of cells in living animals using the combined bioluminescence, fluorescence, and PET imaging techniques.

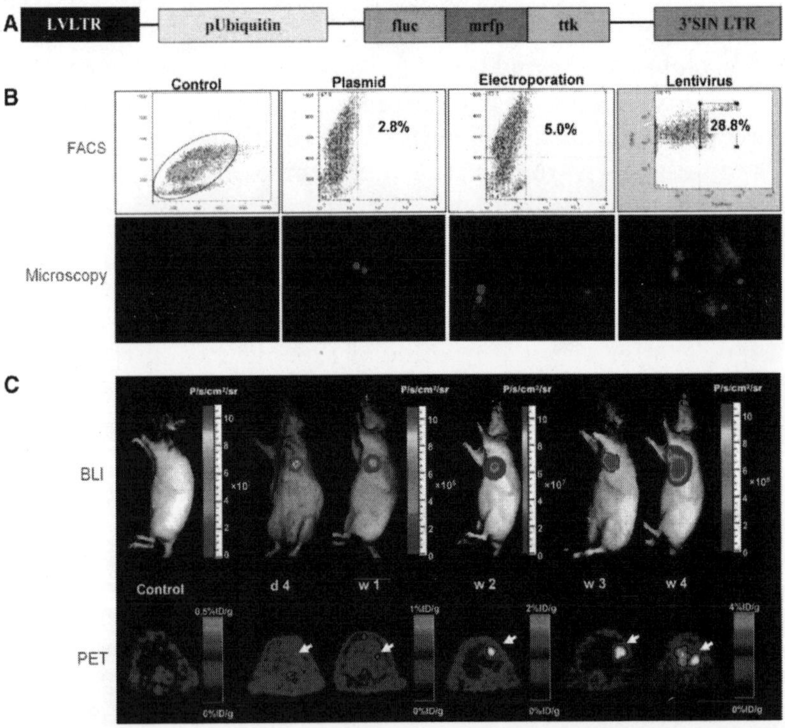

A proof-of-principle study for lentivirus-transduced murine embryonic stem cells (mESCs) that stably express this triple-fusion reporter gene was performed (Cao et al. 2006). It was shown that this molecular imaging platform can be used to monitor the kinetics of stem cell survival, proliferation, migration, and ablation of teratoma sites. The fluorescence feature of the reporter gene was used for cell sorting and microscopy studies, whereas both BLI and PET were used for longitudinal noninvasive monitoring of the transplanted mESCs.

Further development of novel imaging techniques will contribute important insights into the biology and the physiology of transplanted stem cells, leading to significant potential clinical applications for years to come.

Multimodality imaging of mESC survival, proliferation, and migration after cardiac delivery with a triple-fusion reporter gene. (*A*) Schema of the triple-fusion reporter gene. (*B*) Fluorescence-activated cell sorting (FACS) histograms of mESCs at 48 h after transduction with plasmid lipofectamine, electroporation, and lentivirus carrying the triple-fusion reporter gene. (*C*) Noninvasive imaging of transplanted mESCs with BLI and PET. Radiance is given in photons/sec per cm^2/steradian.

Radiolabelled luciferase substrates have been explored for multimodality imaging of gene expression. In one report, a few ^{11}C-labelled D-luciferin analogs were synthesized (Wang et al. 2006). PET studies showed a low retention of the ^{11}C label at 45 min post-injection in luciferase-expressing tumors, whereas BLI with the unlabelled substrate D-luciferin and the radiolabelled analogs gave tumor signal within a few minutes of photon counting. Because D-luciferin and the radiolabelled analogs are substrates of Fluc, with a relatively high turnover rate, this is likely responsible for the poor retention of the tracers in the tumor. In addition, the required injected doses for these two imaging modalities also differ by several orders of magnitude. Therefore, although this is an interesting venue for exploring multimodality imaging, implementing it for optimal performance in animal studies is very challenging.

6

Nucleic Acid Based Encapsulations for Cancer Diagnostics and Drug Delivery

Nanoengineering exploits the interactions of materials at the nanometre scale to create functional nanostructures. It relies on the precise organisation of nanomaterials to achieve unique functionality. There are no interactions more elegant than those governing nucleic acids via Watson–Crick base-pairing rules. The infinite combinations of DNA/RNA base pairs and their remarkable molecular recognition capability can give rise to interesting nanostructures that are only limited by our imagination. Over the past years, creative assembly of nucleic acids has fashioned a plethora of two-dimensional and three-dimensional nanostructures with precisely controlled size, shape and spatial functionalization. These nanostructures have been precisely patterned with molecules, proteins and gold nanoparticles for the observation of chemical reactions at the single molecule level, activation of enzymatic cascade and novel modality of photonic detection, respectively. Recently, they have also been engineered to encapsulate and release bioactive agents in a stimulus-responsive manner for therapeutic applications. The future of nucleic acid-based nanoengineering is bright and exciting.

Nanotechnology is one of the most popular areas of scientific research, especially with regard to medical applications. We've already discussed some of the new detection methods that should bring about cheaper, faster and less invasive cancer diagnoses. But once the diagnosis occurs, there's still the prospect of surgery, chemotherapy or radiation treatment to destroy the cancer. Unfortunately, these treatments can carry serious side effects. Chemotherapy can cause a variety of ailments, including hair loss, digestive problems, nausea, lack of energy and mouth ulcers.

But nanotechnologists think they have an answer for treatment as well, and it comes in the form of targeted drug therapies. If scientists can load their cancer-detecting gold nanoparticles with anticancer drugs, they could attack the cancer exactly where it lives. Such a treatment means fewer side effects and less medication used. Nanoparticles also carry the potential for targeted *and* time-release drugs. A potent dose of drugs could be delivered to a specific area but engineered to release over a planned period to ensure maximum effectiveness and the patient's safety.

These treatments aim to take advantage of the power of nanotechnology and the voracious tendencies of cancer cells, which feast on everything in sight, including drug-laden nanoparticles. One experiment of this type used modified bacteria cells that were 20 percent the size of normal cells. These cells were equipped with antibodies that latched onto cancer cells before releasing the anticancer drugs they contained.

Another used nanoparticles as a companion to other treatments. These particles were sucked up by cancer cells and the cells were then heated with a magnetic field to weaken them. The weakened cancer cells were then much more susceptible to chemotherapy.

It may sound odd, but the dye in your blue jeans or your ballpoint pen has also been paired with gold nanoparticles to fight cancer. This dye, known as phthalocyanine, reacts with light. The nanoparticles take the dye directly to cancer cells while normal cells reject the dye. Once the particles are inside, scientists "activate" them with light to destroy the cancer. Similar therapies have existed to treat skin cancers with light-activated dye, but scientists are now working to use nanoparticles and dye to treat tumors deep in the body.

From manufacturing to medicine to many types of scientific research, nanoparticles are now rather common, but some scientists have voiced concerns about their negative health effects. Nanoparticles' small size allows them to infiltrate almost anywhere. That's great for cancer treatment but potentially harmful to healthy cells and DNA. There are also questions about how to dispose of nanoparticles used in manufacturing or other processes. Special disposal techniques are needed to prevent harmful particles from ending up in the water supply or in the general environment, where they'd be impossible to track.

Gold nanoparticles are a popular choice for medical research, diagnostic testing and cancer treatment, but there are numerous types of nanoparticles in use and in development. Bill Hammack, a professor of chemical engineering at the University of Illinois, warned that nanoparticles are "technologically sweet". In other words, scientists are so wrapped up in what they can do, they're not asking if they should do it. The Food and Drug Administration has a task force on nanotechnology, but as of yet, the government has exerted little oversight or regulation.

Nanorobotics

Nanorobotics is the emerging technology field creating machines or robots whose components are at or close to the scale of a nanometre (10^{-9} metres). More specifically, nanorobotics refers to the nanotechnology engineering discipline of designing and building nanorobots, with devices ranging in size from 0.1–10 micrometres and constructed of nanoscale or molecular components. The names nanobots, nanoids, nanites, nanomachines or nanomites have also been used to describe these devices currently under research and development.

Nanomachines are largely in the research-and-development phase, but some primitive molecular machines and nanomotors have been tested. An example is a sensor having a switch approximately 1.5 nanometres across, capable of counting specific molecules in a chemical sample. The first useful applications of nanomachines might be in medical technology, which could be used to identify and destroy cancer cells. Another potential application is the detection of toxic chemicals, and the measurement of their concentrations, in the environment. Rice University has demonstrated a single-molecule car developed by a chemical process and including buckyballs for wheels. It is actuated by controlling the environmental temperature and by positioning a scanning tunneling microscope tip.

Another definition is a robot that allows precision interactions with nanoscale objects, or can manipulate with nanoscale resolution. Such devices are more related to microscopy or scanning probe microscopy, instead of the description of nanorobots as molecular machine. Following the microscopy definition even a large apparatus such as an atomic force microscope can be considered a nanorobotic instrument when configured to perform nanomanipulation. For this perspective, macroscale robots or microrobots that can move with nanoscale precision can also be considered nanorobots.

According to Richard Feynman, it was his former graduate student and collaborator Albert Hibbs who originally suggested to him (circa 1959) the idea of a *medical* use for Feynman's theoretical micromachines. Hibbs suggested that certain repair machines might one day be reduced in size to the point that it would, in theory, be possible to (as Feynman put it) "swallow the doctor". The idea was incorporated into Feynman's 1959 essay *There's Plenty of Room at the Bottom*. Since nanorobots would be microscopic in size, it would probably be necessary for very large numbers of them to work together to perform microscopic and macroscopic tasks. These nanorobot swarms, both those incapable of replication (as in utility fog) and those capable of unconstrained replication in the natural environment (as in grey goo and its less common variants), are found in many science fiction stories, such as the Borg nanoprobes in *Star Trek* and The Outer Limits episode The New Breed.

Some proponents of nanorobotics, in reaction to the grey goo scenarios that they earlier helped to propagate, hold the view that nanorobots capable of replication outside of a restricted factory environment do not form a necessary part of a purported productive nanotechnology, and that the process of self-replication, if it were ever to be developed, could be made inherently safe. They further assert that their current plans for developing and using molecular manufacturing do not in fact include free-foraging replicators.

The most detailed theoretical discussion of nanorobotics, including specific design issues such as sensing, power communication, navigation, manipulation, locomotion, and onboard computation, has been presented in the medical context of nanomedicine by Robert Freitas. Some of these discussions remain at the level of unbuildable generality and do not approach the level of detailed engineering.

Approaches

Biochip

The joint use of nanoelectronics, photolithography, and new biomaterials provides a possible approach to manufacturing nanorobots for common medical applications, such as for surgical instrumentation, diagnosis and drug delivery. This method for manufacturing on nanotechnology scale is currently in use in the electronics industry. So, practical nanorobots should be integrated as nanoelectronics devices, which will allow tele-operation and advanced capabilities for medical instrumentation.

Nubots

Nubot is an abbreviation for "nucleic acid robot." Nubots are organic molecular machines at the nanoscale. DNA structure can provide means to assemble 2D and 3D nanomechanical devices. DNA based machines can be activated using small molecules, proteins and other molecules of DNA. Biological circuit gates based on DNA materials have been engineered as molecular machines to allow in-vitro drug delivery for targeted health problems. Such material based systems would work most closely to smart biomaterial drug system delivery, while not allowing precise in vivo teleoperation of such engineered prototypes.

Surface-Bound Systems

A number of reports have demonstrated the attachment of synthetic molecular motors to surfaces. These primitive nanomachines have been shown to undergo machine-like motions when confined to the surface of a macroscopic material. The surface anchored motors could potentially be used to move and position nanoscale materials on a surface in the manner of a conveyor belt.

Positional Nanoassembly

Nanofactory Collaboration, founded by Robert Freitas and Ralph Merkle in 2000 and involving 23 researchers from 10 organisations and 4 countries, focuses on developing a practical research agenda specifically aimed at developing positionally-controlled diamond mechanosynthesis and a diamondoid nanofactory that would have the capability of building diamondoid medical nanorobots.

Bacteria-Based

This approach proposes the use of biological microorganisms, like the bacterium *Escherichia coli*. Thus the model uses a flagellum for propulsion purposes. Electromagnetic fields normally control the motion of this kind of biological integrated device. Chemists at the University of Nebraska have created a humidity gauge by fusing a bacteria to a silicone computer chip.

Virus-Based

Some authors also consider virus as a type of nanomachine. A designed modifying HIV Self-Terminating Immuno-Stealth Bloodborne Multi-targeting Retroviral Vector(S.T.I.S.B.M.R.V) is also presented by many authors in not peer-reviewed and some other peer reviewed reference sources as a type of nanodevice that may be used for medicine.

These Nano-Virii repair DNA defects by reintroducing key DNA stabilization enzymes to slow the aging process. The basic principle was based off the journal of virology's research utilising ligase and endonuclease self termination loops controlling the reproduction of the virus.

These Nano-virii are generated by oligonucleotide synthesis along with fractal protoplast fusion DNA recombination in a ligase solution. Although interesting, this approach resembles long established and current vaccine practice, which brings some limitation and possible serious side effects.

Open Technology

A document with a proposal on nanobiotech development using open technology approaches has been addressed to the United Nations General Assembly. According to the document sent to the UN, in the same way that Open Source has in recent years accelerated the development of computer systems, a similar approach should benefit the society at large and accelerate nanorobotics development.

The use of nanobiotechnology should be established as a human heritage for the coming generations, and developed as an open technology based on ethical practices for peaceful purposes. Open technology is stated as a fundamental key for such an aim.

Nanorobot Race

In the same ways that technology development had the space race and nuclear arms race, a race for nanorobots is occurring. There is plenty of ground allowing nanorobots to be included among the emerging technologies. Some of the reasons are that large corporations, such as General Electric, Hewlett-Packard and Northrop Grumman have been recently working in the development and research of nanorobots; surgeons are getting involved and starting to propose ways to apply nanorobots for common medical procedures; universities and research institutes were granted funds by government agencies exceeding $2 billion towards research developing nanodevices for medicine; bankers are also strategically investing with the intent to acquire beforehand rights and royalties on future nanorobots commercialization. Some aspects of nanorobot litigation and related issues linked to monopoly have already arisen. A large number of patents has been granted recently on nanorobots, done mostly for patent agents, companies specialised solely on building patent portfolio, and lawyers.

Potential Applications

Nanomedicine

Potential applications for nanorobotics in medicine include early diagnosis and targeted drug-delivery for cancer, biomedical instrumentation, surgery, pharmacokinetics monitoring of diabetes, and health care.

In such plans, future medical nanotechnology is expected to employ nanorobots injected into the patient to perform work at a cellular level. Such nanorobots intended for use in medicine should be non-replicating, as replication would needlessly increase device complexity, reduce reliability, and interfere with the medical mission.

Nanotechnology provides a wide range of new technologies for developing customized solutions that optimize the delivery of pharmaceutical products. Today, harmful side effects of treatments such as chemotherapy are commonly a result of drug delivery methods that don't pinpoint their intended target cells accurately. Researchers at Harvard and MIT, however, have been able to attach special RNA strands, measuring nearly 10 nm in diameter, to nano-particles, filling them with a chemotherapy drug. These RNA strands are attracted to cancer cells. When the nanoparticle encounters a cancer cell, it adheres to it, and releases the drug into the cancer cell. This directed method of drug delivery has great potential for treating cancer patients while avoiding negative effects (commonly associated with improper drug delivery).

Another useful application of nanorobots is assisting in the repair of tissue cells alongside white blood cells. The recruitment of inflammatory cells or white blood cells (which include neutrophils, lymphocytes, monocytes and mast cells) to the affected area is the first response of tissues to injury. Because of their small size nanorobots could attach themselves to the surface of recruited white cells, to squeeze their way out through the walls of blood vessels and arrive at the injury site, where they can assist in the tissue repair process. Certain substances could possibly be utilised to accelerate the recovery.

The science behind this mechanism is quite complex. Passage of cells across the blood endothelium, a process known as transmigration, is a mechanism involving engagement of cell surface receptors to adhesion molecules, active force exertion and dilation of the vessel walls and physical deformation of the migrating cells. By attaching themselves to migrating inflammatory cells, the robots can in effect "hitch a ride" across the blood vessels, bypassing the need for a complex transmigration mechanism of their own.

The US FDA currently regulates nanotechnology on the basis of size. The FDA also regulates that which acts by chemical means as a drug, and that which acts by physical means as a device. Single molecules can also be used as Turing machines, like their larger paper tape counterparts, capable of universal computation and exerting physical (or chemical) forces as a result of that computation. Safety systems are being developed so that if a drug payload were to be accidentally released, the payload would either be inert or another drug would be then released to counteract the first. Toxicological testing becomes convolved with software validation in such circumstances. With new advances in nanotechnology these small devices are being created with the ability to self-regulate and be 'smarter' than previous generations. As nanotechnology becomes more complex, how will regulatory agencies distinguish a drug from a device? Drug molecules must undergo slower and more expensive testing (for example, preclinical toxicological testing) than devices, and the regulatory pathways for devices are simpler than for drugs. Perhaps smartness, if smart enough, will someday be used to justify a device classification for a single molecule nanomachine. Devices are generally approved more quickly than drugs, so device classification could be beneficial to patients and manufacturers.

Integration of Nanodevices in the Body

Molecular manufacturing (MM) will impact the practice of medicine in many ways. Medicine is highly complex, so it will take some time for the full benefits to be achieved, but many benefits will occur almost immediately. The tools of medicine will become cheaper and more powerful. Research and diagnosis will be far more efficient, allowing rapid response to new diseases, including engineered diseases. Small, cheap, numerous sensors, computers, and other implantable devices may allow continuous health monitoring and semi-automated treatment. Several new kinds of treatment will become possible.

As the practice of medicine becomes cheaper and less uncertain, it can become available to more people.

Surgical and diagnostic tools will be elegant and cheap	Medicine, especially medical research, demands cutting-edge, high-tech tools. These are naturally expensive to manufacture, especially if they must be kept sterile. With a molecular manufacturing system, the cost of production is unrelated to the complexity of the product. Design and testing will still be costly, but once designed, tools can be manufactured in quantity. The incredibly small component size will allow new kinds of tools: for example, a complete surgical robot can be built smaller than a hypodermic needle, and a chemical sensor can be small enough to fit inside a living cell. Because the human body is so complex, accurate knowledge of its state requires gathering large amounts of data. The small size and low cost of nano-built sensors will allow hundreds or thousands of them to be used for routine diagnosis, whereas today only a few data points can be gathered. Integrated sampling and analysis tools will allow real-time monitoring; there will be no need for a separate "lab" to run the tests.
Research and diagnosis will become more efficient	Medical research has traditionally been a process of trial and error. Make a change, then wait a few hours or days to see its effect on the overall state of health. This required an extremely conservative approach, as medical techniques had to evolve one step at a time. With real-time monitoring of the body's systems, it will be possible to detect undesired effects far earlier, allowing a more aggressive and experimental approach to treatment. Researchers will be able to gather far more data and process it with computers millions of times more powerful. The result will be a detailed model of the body's systems and processes, and the ability to predict the effects of any disease or treatment. Diagnosis will also be far easier and more informative. It will be possible to build thousands of diagnostic tests, including invasive tests and imaging tests, into a single, cheap, hand-held device. A variety of single-molecule detection technologies will be available even with early MM. Trustworthy diagnosis will make medicine far more efficient, and also reduce the risk of malpractice (and thus liability insurance).

Contd...

Small medical devices can be implanted permanently	Today, only a few medical devices are implanted permanently. Surgery is always undesirable, and not much functionality can be packed into a device small enough to wear inside the body. Nano-built devices will be far more efficient and compact. As MM technologies gain the ability to synthesize chemicals other than diamond, implantable devices will be able to continuously sense and adjust the body's chemical balance, in the bloodstream or in specific tissues. Even before then, implanted sensors will be valuable in acquiring a continuous record of the person's state of health. This will allow more sensitive adjustment of the body's state, and earlier detection of problems.
More medical problems will be prevented	Many medical problems are preventable. Some are acquired from the environment, including poisoning, some cancers, and almost all infectious disease. Widespread monitoring of health and the environment will allow detection of the source of such problems before they can injure people. Improved infrastructure such as water filtration will also help to reduce environmentally-acquired disease. Other diseases are related to lifestyle. Current lifestyle advice is difficult to follow and is not always accurate. Better research will greatly improve our understanding of cause and effect, allowing us to live more healthy lifestyles with far less effort. Finally, some problems accumulate over time, and early detection and treatment can correct the problem before it turns into a full-blown disease.
New diseases will be stopped quickly	New diseases continue to be a threat to the human race. Naturally occurring diseases could be far worse than SARS, and an engineered disease could conceivably wipe out most of the human race. It will be increasingly important to have a technology base that can detect new diseases even before symptoms appear, and create a cure in a matter of days. MM will enable just such a rapid response. With complete genomes and proteomes for humans and for all known pathogens, plus cheap, highly parallel DNA and protein analysis and sufficient computer resources, it will be possible to spot any new pathogen almost immediately. Curing a new infectious disease will require some method of detecting and stopping the pathogen. Robert Freitas has described over a dozen nanotechnological ways to disable or destroy pathogens.

Contd...

Diagnosis and treatment may be semi-automated	The practice of medicine today involves a lot of uncertainty. Doctors must guess what condition a patient has, and further guess how best to treat it without upsetting the rest of the body's systems. By contrast, when pathogens and chemical imbalances can be directly detected, many conditions will be treatable with no uncertainty, allowing the use of computer-selected treatment in common cases. This may further reduce the cost of medical care, although doctors, regulatory agencies, or the patients themselves may resist the practice initially.
Health will improve and lifespans increase	Health improvement and life extension do not depend directly on molecular manufacturing, but it will certainly make them accessible to more people. Any treatment that can be automated can be applied to any number of people at low cost. Efficient research will speed the development of cures for complex problems such as cancer and aging. New therapeutic techniques will allow the treatment of more types of diseases.
MM will facilitate genetic therapy.	Genetic therapy holds great promise for treating several serious health problems. However, the current state of the art can also cause problems, including cancer. Eventually, we may hope that MM will be able to directly edit the DNA of living cells in the body. But even without that level of sophistication, massively parallel scanning may enable the sorting of cells modified outside the body. The ability to inject only non-cancerous cells would make some kinds of genetic therapy much safer. Microsurgical techniques could allow the implantation of modified cells directly into the target tissues.
Some organs will be replaceable.	Many organs in the body perform fairly simple functions. Already, sophisticated machinery can replace lung function for hours, heart function for months, and kidney function for years. Since MM can build machines smaller than cells, many other organs will be candidates for replacement or augmentation, including skin, muscles, various digestive organs, and some sensory functions.

Contd...

Systems can be individually improved.	The body is made of a large number of interacting systems. The blood circulates chemicals all through the body, making each system interdependent with the others. Small, implanted devices will allow the systems to be decoupled and controlled independently to some degree. For example, it may be desirable for the brain to receive more, or less, adrenaline than the muscles. This capability of "heterostasis" may be useful in cases of trauma and disease, or for long-term health maintenance.

Heart Nanotechnology

Heart nanotechnology is the "Engineering of functional systems at the molecular scale".

Nanomedicine that deals with helping the heart is really starting to take off and gain in popularity compared to most of the other fields that nanomedicine currently has to offer. There are several heart problems that could potentially be used to treat heart disease in the near future.

Examples

It should hopefully be able to treat heart valves that are defective; and detect and treat arterial plaque in the heart ("Nanotechnology Made Clear"). Nanomedicine should be able to help heal the hearts of people that have already been victims of heart disease and heart attacks.

On the other hand it will also play a key role in finding people with a high risk of having heart disease, and will be able to help prevent heart attacks from happening in the first place. Nanotechnology of the heart is a lot less invasive than surgery because everything is occurring at a minuscule level in the body compared to relatively large tissues that are dealt with in surgery. With our technology today, heart surgeries are performed to treat the damaged heart tissue that resulted from a heart attack. This is a major surgery that usually takes a couple of months to recover from.

During this period, patients are extremely limited in the activities that they can do. This long recovery process is an inconvenience to the patients, and with the growth of medicine it most likely won't be very long before a more efficient method for treating heart attack patients will be developed and used.

The method that is the frontrunner to replace major heart surgery is the use of nanotechnology. There are a couple alternate ways to heart surgery that nanotechnology will potentially be able to offer in the future.

Alternatives to Surgery

With people that have heart disease or that have suffered a heart attack, their hearts are often damaged and weakened. The more minor forms of heart failure do not require surgery and are often treated with medications. The use of nanotechnology on treating damaged hearts, will not replace these milder heart problems, but rather the more serious heart problems that currently require surgery or sometimes even heart transplants.

Heart Repair

A group of engineers, doctors and materials scientists at MIT and Children's Hospital Boston have teamed together and are starting the movement of finding a way to use nanotechnology to strengthen the weakened heart tissue. The first method uses nanotechnology combined with tissue engineering, and gold nanowires are placed and woven into the damaged parts of the heart, essentially replacing the non-functioning or dead tissues.

===Tissue regeneration===The other approach would potentially use minuscule nanoparticles that would travel through the body and find dying heart tissue. The nanoparticles would be carrying objects such as "stem cells, growth factors, drugs and other therapeutic compounds,". Then the nanoparticles would release the compounds and inject them into the damaged heart tissue. This would theoretically lead to the regeneration of the tissue.

Heart Repair Difficulties

Being able to fix cardiac tissue that has been damaged from a heart attack or heart disease is not very simple and it is one of the major challenges today in the field of tissue engineering. This is because heart cells are not the easiest objects to create in a lab. It takes an enormous amount of special care and work to develop the cells so that they beat in sync with one another. Even after the heart cells have finally been made, it is also a large task to insert the cells into the inoperable parts of the heart and to get them working in unison with the tissues that were still working properly.

Heart Patches

There have been several successful examples of this with the use of a "stem-cell- based heart patch developed by Duke University researchers". The biomaterials that make up the patch are usually made of either biological polymers like alginate or synthetic polymers such as polylactic acid. These materials are good at organising the cells into functioning tissues; however they act as insulators and are poor conductors of electricity, which is a major problem especially in the heart. Since the electrical signals that are sent between calcium ions are what control when the cardiomyocytes of the heart contract, which makes the heart beat, the stem-cell heart patch is not very efficient and not as effective as doctors would like it to be. The results of the patch not being very conductive is that the cells are not able to attain a smooth, continuous beat throughout the entire tissue containing the stem cells. This results in the heart not functioning properly, which in turn could mean that more heart problems might arise due to the implanting of the stem cells.

Tissue Scaffolds

Recently there have been some new developments in the field of nanotechnology that will be more efficient than the poorly conducting stem-cell-based patch. Scientists and researchers found a way for these stem cell patches (also known as tissue scaffolds) to be conductive and therefore become exponentially more effective. They found that by growing gold nanowires into and through the patches, they were able to greatly increase the electrical conductivity. The nanowires are thicker than the original scaffold and the cells are better organised as well. There is also an increase in production of the proteins needed for muscle calcium binding and contraction.

The gold nanowires poke through the stem cell's scaffolding material, which strengthens the electrical communication between surrounding heart cells. Without the nanowires, the stem cell patches produced a minute current and the cells would only beat in small clusters at the stimulation origin. With the nanowires, the cells seem to contract together even when they are clustered far away from the source of stimulation.

The use of gold nanowires with the stem cell heart patches is still a relatively new concept and it will probably be awhile before they will be used in humans. It is hoped that the nanowires will be tested in live animals in the near future.

Nanoparticles

Another way that nanotechnology will potentially be used to help fix damaged heart tissues is through the use of guided nanoparticle "missiles". These nanoparticles can cling to and attach to artery walls and secrete medicine at a slow rate. The particles, known as nanoburrs due to the fact that they are coated with little protein fragments that stick to and target certain proteins. The nanoburrs can be made to release the drug that is attached to them over the course of several days.

They are unique compared to regular drugs because they can find the particular damaged tissue, attach to it, and release the drug payload that is attached to it. What happens is the nanoburrs are targeted to a certain structure, known as the basement membrane; this membrane lines the arterial walls and is only present if the area is damaged. Nanoburrs could be able to carry drugs that are effective in treating the heart, and also potentially carry stem cells to help regenerate the damaged heart tissue.

Composition

The particles are made up of three different layers and are sixty nanometres in diameter. The outer layer is a coating of polymer called PEG, and its job is to protect the drug from disintegrating while it is travelling through the body. The middle layer consists of a fatty substance and the inner core contains the actual drug along with a polymer chain, which controls the amount of time it will take before the drug is released.

Research

In a study done on rats, the nanoparticles were injected directly into the rat's tail and they still were able to reach the desired target (the left carotid artery) at a rate that was twice the amount of the non-targeted nanoparticles. Because the particles can deliver drugs over a long period of time, and can be injected intravenously, the patients would not need to have multiple repeated injections, or invasive surgeries on the heart which would be a lot more convenient. The only downside to this is that the existing delivery approaches are invasive, requiring either a direct injection into the heart, catheter procedures, or surgical implants. There is no question, however, that the future of heart repairs and heart disease/attack prevention will definitely involve the use of nanotechnology in some way.

Polyketal Nanoparticles

Composition

Polyketal nanoparticles are pH-sensitive, hydrophobic nanoparticles formulated from poly(1-4-phenyleneacetone dimethylene ketal). They are an acid-sensitive vehicle of drug delivery, specifically designed for targeting the environments of tumors, phagosomes, and inflammatory tissue. In such acidic environments, these nanoparticles undergo accelerated hydrolysis into low molecular weight hydrophilic compounds, consequently releasing their therapeutic contents at a faster rate. Unlike polyester-based nanoparticles, polyketal nanoparticles do not generate acidic degradation products following hydrolysis

Use in Myocardial Infarction

Post-myocardial infarction, inflammatory leukocytes invade the myocardium. Leukocytes contain high amounts of Nicotinamide adenine dinucleotide phosphate (NADPH) and Nox2. Nox2 and NADPH oxidase combine to act as a major source of cardiac superoxide production, which in excess can lead to myocyte hypertrophy, apoptosis, fibrosis, and increased matrix metalloproteinase-2 expression. In a mouse-model study by Somasuntharam et al. 2013, polyketal nanoparticles were used as a delivery vehicle for siRNA to target and inhibit Nox2 in the infarcted heart. Following intramyocardial injection in vivo, Nox2-siRNA nanoparticles prevented upregulation of Nox2-NADPH oxidase, and improved fractional shortening. When taken up by macrophages in the myocardium following a MI, the nanoparticles degraded in the acidic environment of the endosomes/phagosomes, releasing Nox2-specific siRNA into the cytoplasm.

Polyketal nanoparticles have also been used in the infarcted mouse heart to prevent ischemia-reperfusion injury caused by reactive oxygen species (ROS). Levels of the antioxidant Cu/Zn-superoxide dismutase (SOD1), which scavenges harmful ROS, decrease following MI. SOD1-enacapsulated polyketal nanoparticles are able to scavenge reperfusion-injury induced ROS. Furthermore, this treatment improved fractional shortening, suggesting the benefit of targeted delivery by polyketals.

One of the key advantages of polyketal use is that they do not exacerbate the inflammatory response, even when administered at concentrations exceeding therapeutic limits. In contrast to commonly-used poly(lactic-co-glycolic acid) (PLGA) nanoparticles, polyketal nanoparticle administration in mice instigates little recruitment of inflammatory cells. Additionally, intramuscular injection of polyketals into the leg of rats shows no significant increases in inflammatory cytokines such as IL-6, IL-1ß, TNF-á and IL-12.

Carbon Nanotubes: Prospects for Tissue Engineering, Therapeutics, and Medical Diagnostics

Carbon nanotubes (CNTs) have been explored in recent years as nanoscale tools for intracellular delivery, due to their ability to be internalised within cells in a variety of different mechanisms depending on their physicochemical characteristics. In addition, their large surface area offers an attractive platform for further chemical derivatization with a variety of molecules. Moreover, they have unique electrical, thermal and spectroscopic properties which further extend their potential in the detection, monitoring and therapy of diseas.

Carbon nanotubes (CNTs) are very prevalent in today's world of medical research and are being highly researched in the fields of efficient drug delivery and biosensing methods for disease treatment and health monitoring. Carbon nanotube technology has shown to have the potential to alter drug delivery and biosensing methods for the better, and thus, carbon nanotubes have recently garnered interest in the field of medicine.

The use of CNTs in drug delivery and biosensing technology has the potential to revolutionalize medicine. Functionalization of single-walled nanotubes (SWNTs) has proven to enhance solubility and allow for efficient tumor targeting/drug delivery. It prevents SWNTs from being cytotoxic and altering the function of immune cells.

Cancer, a group of diseases in which cells grow and divide abnormally, is one of the primary diseases being looked at with regards to how it responds to CNT drug delivery. Current cancer therapy primarily involves surgery, radiation therapy, and chemotherapy. These methods of treatment are usually painful and kill normal cells in addition to producing adverse side effects. CNTs as drug delivery vehicles have shown potential in targeting specific cancer cells with a dosage lower than conventional drugs used, that is just as effective in killing the cells, however does not harm healthy cells and significantly reduces side effects. Current blood glucose monitoring methods by patients suffering from diabetes are normally invasive and often painful. For example, one method involves a continuous glucose sensor integrated into a small needle which must be inserted under the skin to monitor glucose levels every few days. Another method involves glucose monitoring strips to which blood must be applied. These methods are not only invasive but they can also yield inaccurate results. It was shown that 70 percent of glucose readings obtained by continuous glucose sensors differed by 10 percent or more and 7 percent differed by over 50 percent. The high electrochemically accessible surface area, high electrical conductivity and useful structural properties have demonstrated the potential use of single-walled nanotubes (SWNTs) and multi-walled nanotubes (MWNTs) in highly sensitive noninvasive glucose detectors.

CNT Properties

CNTs have several unique chemical, size, optical, electrical and structural properties that make them attractive as drug delivery and biosensing platforms for the treatment of various diseases and the noninvasive monitoring of blood levels and other chemical properties of the human body, respectively.

Electrical and Structural

Carbon nanotubes can be metallic or semiconducting depending on their structure. This is due to the symmetry and unique electronic structure of graphene. For a given (n,m) nanotube, if n = m, the nanotube is metallic; if n “ m is a multiple of 3, then the nanotube is semiconducting with a very small band gap, otherwise the nanotube is a moderate semiconductor. Thus all armchair (n=m) nanotubes are metallic, and nanotubes (5,0), (6,4), (9,1), etc. are semiconducting. Thus, some nanotubes have conductivities higher than that of copper, while others behave more like silicon.

Dimensional

Due to their nanoscale dimensions, electron transport in carbon nanotubes will take place through quantum effects and will only propagate along the axis of the tube. These electrical and structural properties best serve CNTs as far as biosensing is concerned because current changes in the CNTs can signify specific biological entities they are designed to detect. The fact that CNTs are small (nm scale) allows them to deliver smaller doses of drugs to specific disease cells in the body thus reducing side effects and harm to healthy cells unlike conventional drugs, whilst improving disease cell targeting efficiency.

Chemical

CNTs have been observed to have enhanced solubility when functionalized with lipids which would make their movement through the human body easier and would also reduce the risk of blockage of vital body organ pathways. As far as optical properties are concerned CNTs have been shown to exhibit strong optical absorbance in certain spectral windows such as NIR (near-infrared) light and when functionalized with tumor cell specific binding entities have allowed the selective destruction of disease (e.g. cancer) cells with NIR in drug delivery applications.

CNTs in Drug Delivery and Cancer Therapy

Drug delivery is a rapidly growing area that is now taking advantage of nanotube technology. Systems being used currently for drug delivery include dendrimers, polymers, and liposomes, but carbon nanotubes present the opportunity to work with effective structures that have high drug loading capacities and good cell penetration qualities. These nanotubes function with a larger inner volume to be used as the drug container, large aspect ratios for numerous functionalization attachments, and the ability to be readily taken up by the cell. Because of their tube structure, carbon nanotubes can be made with or without end caps, meaning that without end caps the inside where the drug is held would be more accessible. Right now with carbon nanotube drug delivery systems, problems arise like the lack of solubility, clumping occurrences, and half-life. However, these are all issues that are currently being addressed and altered for further advancements in the carbon nanotube field. The advantages of carbon nanotubes as nanovectors for drug delivery remain where cell uptake of these structures was demonstrated efficiently where the effects were prominent, showing the particular nanotubes can be less harmful as nenovehicles for drugs. Also, drug encapsulation has been shown to enhance water dispersibility, better bioavailability, and reduced toxicity. Encapsulation of molecules also provides a material storage application as well as protection and controlled release of loaded molecules. All of these result in a good drug delivery basis where further research and understanding could improve upon numerous other advancements, like increased water solubility, decreased toxicity, sustained half-life, increased cell penetration and uptake, all of which are currently novel but undeveloped ideas.

Boron Neutron Capture Therapy

Narayan Hosmane and his co-workers have recently developed a new approach to Boron Neutron Capture Therapy in the treatment of cancer using substituted Carborane-Appended Water-Soluble single-wall carbon nanotubes. Substituted C2B10 carborane cages were successfully attached to the side walls of single wall carbon nanotubes (SWCNTs) via nitrene cycloaddition. The decapitations of these C2B10 carborane cages, with the appended SWCNTs intact, were accomplished by the reaction with sodium hydroxide in refluxing ethanol. During base reflux, the three-membered ring formed by the nitrene and SWCNT was opened to produce water-soluble SWCNTs in which the side walls were functionalized by both substituted nido-C2B9 carborane units and ethoxide moieties. All new compounds were characterized by EA, SEM, TEM, UV, NMR, and IR spectra and chemical analyses. Selected tissue distribution studies on one of these nanotubes, {([Na+][1-Me-2-((CH2)4NH-)-1,2-C2B9H10][OEt])n(SWCNT)} (Va), showed that the boron atoms are concentrated more in tumors cells than in blood and other organs, making it an attractive nanovehicle for the delivery of boron to tumor cells for an effective boron neutron capture therapy in the treatment of cancer.

Selective Cancer Cell Destruction

Carbon nanotubes can be used as multifunctional biological transporters and near-infrared agents for selective cancer cell destruction. Biological systems are known to be highly transparent to 700- to 1,100-nm near-infrared (NIR) light. Researchers showed that the strong optical absorbance of single-walled carbon nanotubes (SWNTs) in this special spectral window, an intrinsic property of SWNTs, can be used for optical stimulation of nanotubes inside living cells to afford multifunctional nanotube biological transporters. They used oligonucleotides transported inside living Hela cells by nanotubes. The oligonucleotides translocated into the cell nucleus upon endosomal rupture triggered by NIR laser pulses. Continuous NIR radiation caused cell death because of excessive local heating of SWNT in vitro. Selective cancer cell destruction was achieved by functionalization of SWNT with a folate moiety, selective internalization of SWNTs inside cells labelled with folate receptor tumor markers, and NIR-triggered cell death, without harming receptor-free normal cells. Thus, the transporting capabilities of carbon nanotubes combined with suitable functionalization chemistry and their intrinsic optical properties can lead to new classes of novel nanomaterials for drug delivery and cancer therapy.

Tumor Targeting

Research has been conducted on in vivo biodistribution and highly efficient tumor targeting of carbon nanotubes in mice for cancer therapy. Investigations are being done on the biodistribution of radio-labelled SWNTs in mice by in vivo positron emission tomography (PET), ex vivo biodistribution and Raman spectroscopy. It was found that SWNTs that are functionalized with phospholipids bearing polyethylene glycol (PEG) are surprisingly stable *in vivo*. The effect of PEG chain length on the biodistribution and circulation of the SWNTs was studied. Effectively PEGylated SWNTs exhibited relatively long blood circulation times and low uptake by the reticuloendothelial system (RES). Efficient targeting of integrin positive tumor in mice was achieved with SWNTs coated with PEG chains linked to an arginine–glycine–aspartic acid (RGD) peptide. A high tumor accumulation was attributed to the multivalent effect of the SWNTs. The Raman signatures of SWNTs were used to directly probe the presence of nanotubes in mice tissues and confirm the radio-label-based results.

CNTs as Biosensors

CNT Network Bio-Stress Sensors

A single nanotube experiences a change in electrical resistance when experiencing stress or strain. This piezoresistive effect changes the current flow through the nanotube, which can be measured in order to accurately quantify the applied stress. A semi-random positioning of many overlapping nanotubes forms an electrically conducting network composed of many piezoresistive nanotubes. If the variance of the tube lengths and angles are known and controllable during manufacture, an eigensystem approach can be used to determine the expected current flow between any two points in the network. The tube network is embedded within orthopedic plates, clamps, and screws and in bone grafts Bone grafting in order to determine the state of bone healing by measuring the effect of a load on the plate, clamp, screw, or other fixation device attached to the bone. A healed bone will bear most of the load while a yet unhealed bone will defer the load to the fixation device wherein the nanotube network may measure the change in resistivity. Measurement is done wirelessly by electrical induction. This allows the doctor to accurately assess patient healing and also allows the patient to know how much stress the affected area may safely tolerate. Wolff's Law indicates that bone responds positively to safe amounts of stress, which may be necessary for proper healing.

Glucose Detection Biosensors

Carbon nanotube–plasma polymer-based amperometric biosensors for ultrasensitive glucose detection have been fabricated. Two amperometric enzyme biosensors were fabricated. One had single wall nanotubes and the other multi wall nanotubes, however, plasma-polymerized thin films (PPFs) were incorporated into both. A mixture of the enzyme glucose oxidase (GOD) and a CNT film was sandwiched with 10-nm-thick acetonitrile PPFs. A PPF layer was deposited onto a sputtered gold electrode. In order to facilitate the electrochemical communication between the CNT layer and GOD, CNTs were treated with oxygen plasma. The device with single-walled CNTs showed a sensitivity higher than that of multi walled CNTs. The glucose biosensor showed ultrasensitivity (a sensitivity of 40 μA mM-1 cm-2, a correlation coefficient of 0.992, a linear response range of 0.025 –1.9 mM, a detection limit of 6.2 μM at S/N = 3, +0.8V vs Ag/AgCl), and a rapid response (<4 seconds in reaching 95% of maximum response). This high performance is attributed to the fact that CNTs have excellent electrocatalytic activity and enhance electron transfer, and that PPFs and/or the plasma process for CNTs are an enzyme-friendly platform, i.e., a suitable design of the interface between GOD and CNTs.

DNA Detection Biosensors

An aligned carbon nanotube ultrasensitive biosensor for DNA detection was developed. The design and fabrication of the biosensor was based on aligned single wall carbon nanotubes (SWCNTs) with integrated single-strand DNAs (ssDNA). The fabricated ultra-sensitive biosensor provided label-free real-time electronic detection of DNA hybridization between surface immobilized ssDNA and target ssDNA. Hybridization kinetics between complementary and target ssDNA nucleotide base pairs resulted in a local charge generation between base pairs that was injected into the SWCNTs resulting in a detectable change in SWCNT electrical conductance. This conductance change was amplified electrically through the integration of the functionalized SWCNTs as the semi-conductive channel in a silicon-silicon oxide based field effect transistor (FET). Based on previous Langmuir DNA kinetics calculations, the projected sensitivity level of the SWCNT-DNA sensor was considerably higher than traditional fluorescent and hybridization assays.

CNT Modified Electrode Biosensors

A microbial biosensor based on carbon nanotube (CNT) modified electrodes was developed. Pseudomonas putida DSM 50026 cells were used as the biological component and the measurement was based on the respiratory activity of the cells estimated from electrochemical measurements. The cells were immobilized on carbon nanotube (CNT) modified carbon paste electrodes (CPE) by means of a redox osmium polymer. The osmium polymer efficiently shuttled electrons between redox enzymes located in the cell wall of the cells and promoted a stable binding to the electrode surface. The effect of varying the amounts of CNT and osmium polymer, on the response to glucose was investigated to find the optimum composition of the sensor. The effects of pH and temperature were also examined. After the optimisation studies, the system was characterised by using glucose as a substrate. Moreover, the microbial biosensor was also prepared by using phenol adapted bacteria and then, calibrated to phenol. After that, it was applied for phenol detection in an artificial waste water sample. The study found that whole cell P. putida biosensors using Os-redox polymers could be good alternatives for the analysis of different substrates such as glucose as well as xenobiotics in the absence of oxygen with high sensitivity because of the fast electron collection efficiency between the Os-redox polymer and the bacterial cells. The use of optimum amounts of CNTs and the Os redox mediator provided better sensor sensitivity by promoting the electron transfer within the structure of the biosensor. The main disadvantages were the high surface area of CNTs that increased the background current and the diffusion problem of electrons that occurred due to overlapping of the diffusion layers formed at closely spaced CNTs in the film. However, these problems could be overcome by optimising the CNT and polymer amounts.

Toxicity Issues

Cytotoxity of Functionalized CNTs

Research shows that functionalized carbon nanotubes are non-cytotoxic and preserve the functionality of primary immune cells. Two types of f-CNTs were prepared, following the 1,3-dipolar cycloaddition reaction (f-CNTs 1 and 2) and the oxidation/amidation treatment (f-CNTs 3 and 4), respectively. Both types of f-CNTs were uptaken by B and T lymphocytes as well as macrophages in vitro, without affecting cell viability. Subsequently, the functionality of the different cells was analyzed carefully. It was discovered that f-CNT 1, which is highly water soluble, did not influence the functional activity of immunoregulatory cells. f-CNT 3, which instead possesses reduced solubility and forms mainly stable water suspensions, preserved lymphocytes' functionality while provoking secretion of proinflammatory cytokines by macrophages. One important thing to note from this study is the fact that certain types of CNTs functionalized with lipids are highly water soluble which would make their movement through the human body easier and would also reduce the risk of blockage of vital body organ pathways thus making them more attractive as drug delivery vehicles.

In Vitro Cytotoxicity

In vitro toxicity of single- and multi-walled carbon nanotubes in human astrocytoma and lung carcinoma cells was investigated. The study was undertaken to characterize the physicochemical properties of single-walled nanotubes (SWNTs), multi-walled nanotubes (MWNTs) and functionalized MW (MW-COOH and MW-NH2), and to assess their cytotoxicity in human astrocytoma D384-cells and lung carcinoma A549-cells, using the MTT assay and calcein/propidium iodide (PI) staining. Both the as-received and the modified nanotubes were characterized by means of thermal analysis (TGA), infrared spectroscopy and atomic force microscopy chiefly to check the degree of functionalization. The cells were exposed to the nanomaterials (0.1–100 μg/ml) for 24, 48 and 72 hours in a medium containing 10% FCS. In D384 cells MTT results revealed a strong cytotoxicity (50%) of SWNTs after 24 hour exposure already at 0.1 μg/ml, without further changes at higher concentrations or longer incubation times. At all time-points MTT metabolism was decreased by 50% by all the other compounds at 10 μg/ml and with no exacerbation at the higher dose. Similar results were obtained with A549 cells. Experiments using calcein/PI staining did not confirm MTT cytotoxicity data neither in D384- nor in A549-cells. The viability of these cells was not affected by any nanotube at any concentration or time of exposure, with the exception of the positive control SiO2. The results suggested the need of a careful examination of carbon nanotubes toxic effects by means of multiple tests to circumvent the possible problem of artifactual results due to the interference of nanomaterials with the dye markers employed.

Cytotoxicity of SWNTs and MWNTs

The cytotoxicity was investigated on healthy alveolar macrophage cells obtained from adult guinea pigs for single-wall nanotubes (SWNTs), multi-wall nanotubes (with diameters ranging from 10 to 20 nm, MWNT10), and fullerene (C60) for comparison purposes. Profound cytotoxicity of SWNTs was observed in alveolar macrophage (AM) after a 6 hour exposure in vitro. The cytotoxicity increased by as high as ~35% when the dosage of SWNTs was increased by 11.30 μg/cm2. No significant toxicity was observed for C60 up to a dose of 226.00 μg/cm2. The cytotoxicity apparently followed a sequence order on a mass basis: SWNTs > MWNT10 > quartz > C60. SWNTs significantly impaired phagocytosis of AM at the low dose of 0.38 μg/cm2, whereas MWNT10 and C60 induced injury only at the high dose of 3.06 μg/cm2. The macrophages exposed to SWNTs or MWNT10 of 3.06 μg/cm2 showed characteristic features of necrosis and degeneration. A sign of apoptotic cell death likely existed. It was concluded from the study that carbon nanomaterials with different geometric structures exhibit quite different cytotoxicity and bioactivity in vitro, although they may not be accurately reflected in the comparative toxicity in vivo.

Future Steps

Further research needs to be done in investigating the long term effects of f-SWNTs in mice and their swift clearance from mice. SWNTs of various lengths should be investigated in the future for elucidating size and shape effects on nanomaterial distribution within tumors to shed more light on the applications of the structural and physical properties of SWNTs for potential therapeutic approaches. Glucose detection can be greatly enhanced using CNTs in a nanometre thick enzyme friendly platform allowing for easy electron transfer between the CNTs and the primary metal electrode and hence allowing for an alternative noninvasive way of monitoring blood glucose. More needs to be done concerning the identification of mechanisms responsible for certain responses in order to design optimal biosensors for the detection of other biological/chemical entities such as DNA, viruses, bacteria and pathogens in addition to glucose.

Carbon nanotubes are still a relatively unexplored area, and the rapidly advancing fields that they are involved in can still be pushed farther. Nanotubes are extremely versatile since they can be included in numerous different fields because of their great material properties. Any amount of improvements can be made to carbon nanotubes through various techniques. For example, it was shown that by electrospinning and plasma-functionalizing single-walled nanotubes, adhesion to surrounding polymer matrices was greatly improved along with the tensile properties of the nanotubes. Also, it is known that most nanotubes are cleared from the body very quickly after being distributed throughout. This decreases the chances of higher toxicity levels in the blood. Many other properties increase the number of options available with carbon nanotubes. The good functionalization of carbon nanotubes allows us to attach a number of groups to the tubes for different systems. Radioactive labels could be attached for use in bioimaging. As mentioned before, fluorescence is already observed in normal carbon nanotubes, but attaching labels allows for a greater imaging window. This labelling capability can also be used for targeting purposes. For example, attaching targeting groups to the carbon nanotubes could open up doors for very specific drug delivery systems. It was shown that carbon nanotubes were used to deliver drugs to specific cancer cells of the epithelium, and this was accomplished efficiently. This targeted killing of cancer cells shows promise for numerous other improvements in cancer therapy and treatment as well as the treatment of various infectious diseases. With targeted nanotubes used for drug delivery, specific cells could be aimed at to take up the carbon nanotubes, as was shown with brain tumor cells. Also with these studies, minimal toxicity was found when multi-walled carbon nanotubes were injected into mice. Carbon nanotubes seem to be a valuable option when considering such applications as drug delivery or bioimaging because they are readily functionalized, display excellent material properties, can be used as imaging agents or sensors, and keep the door open for many future advances.

Antibody-Drug Conjugate

Antibody-drug conjugates or ADCs are a new class of highly potent biopharmaceutical drugs designed as a targeted therapy for the treatment of people with cancer. ADCs are complex molecules composed of an antibody (a whole mAb or an antibody fragment such as a single-chain variable fragment [scFv]) linked, via a stable, chemical, linker with labile bonds, to a biological active cytotoxic (anticancer) payload or drug. Antibody Drug Conjugates are examples of bioconjugates and immunoconjugates.

By combining the unique targeting capabilities of monoclonal antibodies with the cancer-killing ability of cytotoxic drugs, antibody-drug conjugates allow sensitive discrimination between healthy and diseased tissue. This means that, in contrast to traditional chemotherapeutic agents, antibody-drug conjugates target and attack the cancer cell so that healthy cells are less severely affected.

Mechanism of Action

In developing antibody-drug conjugates, an anticancer drug (e.g. a cell toxin or cytotoxin) is coupled to an antibody that specifically targets a certain tumor marker (e.g. a protein that, ideally, is only to be found in or on tumor cells). Antibodies track these proteins down in the body and attach themselves to the surface of cancer cells. The biochemical reaction between the antibody and the target protein (antigen) triggers a signal in the tumor cell, which then absorbs or internalizes the antibody together with the cytotoxin. After the ADC is internalized, the cytotoxic drug is released and kills the cancer. Due to this targeting, ideally the drug has lower side effects and gives a wider therapeutic window than other chemotherapeutic agents. These advantages have led to ADC technologies being featured in many publications, notably the New York Times, as well as numerous scientific journals.

Market Approval

To date, only three ADCs have received market approval. However, after a request from the U.S. Food and Drug Administration (FDA), Pfizer/Wyeth, the developer and marketer of the first ADC to receive marketing approval in 2001 for the treatment of patients with acute myelogenous leukemia (Gemtuzumab ozogamicin, trade name: Mylotarg), withdrew the drug from the market in June 2010. As a result, only two ADC are marketed (2013), including Brentuximab vedotin (trade name: Adcetris, marketed by Seattle Genetics and Millennium/Takeda) and Trastuzumab emtansine (trade name: Kadcyla, marketed by Genentech and Roche).

Brentuximab vedotin was granted accelerated approval by the U.S. Food and Drug Administration (FDA) on August 19, 2011 for relapsed HL and relapsed sALCL and conditional Marketing authorization from the European Medicines Agency in October 2012 for relapsed or refractory HL and relapsed or refractory sALCL. Trastuzumab emtansine (ado-trastuzumab emtansine or T-DM1) was approved in February 2013 for the treatment of people with HER2-positive metastatic breast cancer (mBC) who have received prior treatment with trastuzumab (Herceptin®, Genentech and Roche) and a taxane chemotherapy.

Linker Technology

A stable link between the antibody and cytotoxic (anti-cancer) agent is a crucial aspect of an ADC. Linkers are based on chemical motifs including disulfides, hydrazones or peptides (cleavable), or thioethers (noncleavable) and control the distribution and delivery of the cytotoxic agent to the target cell. Cleavable and noncleavable types of linkers have been proven to be safe in preclinical and clinical trials. Brentuximab vedotin includes an enzyme-sensitive *cleavable linker* that delivers the potent and highly toxic antimicrotubule agent monomethyl auristatin E or MMAE, a synthetic antineoplastic agent, to human specific CD30-positive malignant cells. Because of its high toxicity MMAE, which inhibits cell division by blocking the polymerization of tubulin, cannot be used as a single-agent chemotherapeutic drug. However, the combination of MMAE linked to an anti-CD30 monoclonal antibody (cAC10, a cell membrane protein of the tumor necrosis factor or TNF receptor) proved to be stable in extracellular fluid, cleavable by cathepsin and safe for therapy. Trastuzumab emtansine, the other approved ADC, is a combination of the microtubule-formation inhibitor mertansine (DM-1), a derivative of the Maytansine, and antibody trastuzumab (Herceptin/ Genentech/Roche) attached by a stable, *non-cleavable linker*.

The availability of better and more stable linkers has changed the function of the chemical bond. The type of linker, *cleavable* or *noncleavable*, lends specific properties to the cytotoxic (anti-cancer) drug. For example, a *non-cleavable* linker keeps the drug within the cell. As a result, the entire antibody, linker and cytotoxic (anti-cancer) agent enter the targeted cancer cell where the antibody is degraded to the level of an amino acid. The resulting complex – amino acid, linker and cytotoxic agent – now becomes the active drug. In contrast, *cleavable linkers* are catalyzed by enzymes in the cancer cell where it releases the cytotoxic agent. The difference is that the cytotoxic payload delivered via a cleavable linker can escape from the targeted cell and, in a process called "bystander killing," attack neighbouring cancer cells.

Another type of cleavable linker, currently in development, adds an extra molecule between the cytotoxic drug and the cleavage site. This linker technology allows researchers to create ADCs with more flexibility without worrying about changing cleavage kinetics. Researchers are also developing a new method of peptide cleavage based on Edman degradation, a method of sequencing amino acids in a peptide. Future direction in the development of ADCs also include the development of site-specific conjugation (TDCs) to further improve stability and therapeutic index and á emitting immunoconjugates and antibody-conjugated nanoparticles.

Non-Natural Amino Acids

The first generation of ADCs use linking technologies that conjugate drugs non-selectively to cysteine or lysine residues in the antibody, resulting in a heterogeneous mixture of ADCs. This approach leads to suboptimal safety and efficacy properties and makes optimization of the biological, physical and pharmacological properties of an ADC challenging. The site-specific incorporation of unnatural amino acids into the antibody generates a site for controlled and stable attachment of the drug. This enables the production of homogeneous ADCs with the antibody precisely linked to the drug and controlled ratios of antibody to drug, allowing the selection of a best-in-class ADC. An Escherichia coli-based open cell-free synthesis (OCFS), developed by Stanford Professor James R. Swartz, allows the synthesis of proteins containing site-specifically incorporated non-natural amino acids and has been optimized for predictable high-yield protein synthesis and folding at any scale with straightforward downstream purification processes. The absence of a cell wall allows the addition of non-natural factors to the open system in order to manipulate transcription, translation and folding to provide precise modulation of the protein expression process.

Other Disease Areas

The majority of the ADCs currently under development or in clinical trials are for oncological and hematological indications. This is primarily driven by the availability of monoclonal antibodies targeting various types of cancer. However, some drug developers are also looking to expanding the application of ADCs beyond oncology and hematology to other important disease areas. At this time, these ADCs are only in initial drug discovery or preclinical stage of development.

Thin Film Drug Delivery

Thin-film drug delivery uses a dissolving film or oral drug strip to administer drugs via absorption in the mouth (buccally or sublingually) and/or via the small intestines (enterically). A film is prepared using hydrophilic polymers that rapidly dissolves on the tongue or buccal cavity, delivering the drug to the systemic circulation via dissolution when contact with liquid is made.

Thin-film drug delivery has emerged as an advanced alternative to the traditional tablets, capsules and liquids often associated with prescription and OTC medications. Similar in size, shape and thickness to a postage stamp, thin-film strips are typically designed for oral administration, with the user placing the strip on or under the tongue (sublingual) or along the inside of the cheek (buccal). These drug delivery options allow the medication to bypass the first pass metabolism thereby making the medication more bioavailable. As the strip dissolves, the drug can enter the blood stream enterically, buccally or sublingually. Evaluating the systemic transmucosal drug delivery, the buccal mucosa is the preferred region as compared to the sublingual mucosa.

Different buccal delivery products have been marketed or are proposed for certain diseases like trigeminal neuralgia, Meniere's disease, diabetes, and addiction. There are many commercial non-drug product to use thin films like Meltz Super Thin Mints and Listerine PocketPaks breath freshening strips. Since then, thin-film products for other breath fresheners, as well as a number of cold, flu, anti-snoring and gastrointestinal medications, have entered the marketplace. There are currently several projects in development that will deliver prescription drugs using the thin-film dosage form.

Formulation of oral drug strips involves the application of both aesthetic and performance characteristics such as strip-forming polymers, plasticizers, active pharmaceutical ingredient, sweetening agents, saliva stimulating agent, flavouring agents, colouring agents, stabilizing and thickening agents. From the regulatory perspectives, all excipients used in the formulation of oral drug strips should be approved for use in oral pharmaceutical dosage forms.

Oral Drug Strip Development

Strip Forming Polymers

The polymer employed should be non-toxic, non-irritant and devoid of leachable impurities. It should have good wetting and spreadability property. The polymer should exhibit sufficient peel, shear and tensile strengths. The polymer should be readily available and should not be very expensive. Film obtained should be tough enough so that there won't be any damage while handling or during transportation. Combination of microcrystalline cellulose and maltodextrin has been used to formulate Oral Strips of piroxicam made by hot melt extrusion technique. Pullulan has been the most widely used film former (used in Listerine PocketPak, Benadryl, etc.)

Plasticizer

Plasticizer is a vital ingredient of the OS formulation. It helps to improve the flexibility of the strip and reduces the brittleness of the strip. Plasticizer significantly improves the strip properties by reducing the glass transition temperature of the polymer. Glycerol, Propylene glycol, low molecular weight polyethylene glycols, phthalate derivatives like dimethyl, diethyl and dibutyl phthalate, Citrate derivatives such as tributyl, triethyl, acetyl citrate, triacetin and castor oil are some of the commonly used plasticizer excipients.

Active Pharmaceutical Ingredient

Since the size of the dosage form has limitation, high-dose molecules are difficult to be incorporated in OS. Generally 5%w/w to 30%w/w of active pharmaceutical ingredients can be incorporated in the OS.

Sweeting, Flavouring and Colouring Agent

An important aspect of thin film drug technology is its taste and colour. The sweet taste in formulation is more important in case of pediatric population. Natural sweeteners as well as artificial sweeteners are used to improve the flavour of the mouth dissolving formulations for the flavours changes from individual to individual. Pigments such as titaniumdioxide is incorporated for colouring.

Stabilizing and Thickening Agents

The stabilizing and thickening agents are employed to improve the viscosity and consistency of dispersion or solution of the strip preparation solution or suspension before casting.

Drug content uniformity is a requirement for all dosage forms, particularly those containing low dose highly potent drugs. To uniquely meet this requirement, thin film formulations contain uniform dispersions of drug throughout the whole manufacturing process.

Since this criterion is essential for the quality of the thin film and final pharmaceutical dosage form, the use of Laser Scanning Confocal Microscopy (LSCM) was recommended to follow the manufacturing process.

Commercial Oral Strip Drugs

There are not yet many medications available in a thin film form on the market. Those that are include:

- Zuplenz: The first oral soluble film approved by the FDA as a prescription medication.
- Benadryl (diphenhydramine product, anti-cholinergic anti-histamine used for allergies and as a mild sedative)
- Gas-X (simethicone product for bloating, gas, and gastrointestinal complaint)
- Melatonin PM (hormonal product sold as a "dietary supplement" marketed for insomnia)
- Orajel Kids (benzocaine product for dental pain)
- Suboxone (buprenorphine and naloxone fixed dosage combination product for opioid addiction)
- Subutex (buprenorphine product for opioid addiction)
- Sudafed (phenylephrine or pseudoephedrine product for nasal congestion)
- Suppress
- TheraFlu (combination product of pain reliever, anti-pyretic and decongestant)
- Triaminic (children's anti-tussive product)

On July 2, 2010, Strativa Pharmaceuticals, the proprietary products division of Par Pharmaceutical, received approval from the US FDA for Zuplenz (ondansetron) oral soluble film (OSF) for the prevention of postoperative, highly and moderately emetogenic cancer chemotherapy-induced, and radiotherapy-induced nausea and vomiting.

Oral Strips in Development

An increasing number of film-based therapeutics are in development, including:

- Montelukast indicated for the treatment of asthma and allergy, is being developed for use as a film by MonosolRx.
- Midatech, a company specialising in nanotechnology, is partnering with MonosolRx to create a film-based insulin. (Sachs Associates. 5th Annual European Life Science CEP Forum for Partnering and Investing. March 6-7, 2012. Zurich, Switzerland.)
- Rizatriptan indicated for the treatment of migraine, is being developed for use as a film by MonosolRx.
- MonosolRx is also developing a testosterone film-based therapeutic for the treatment of male hypogonadism. The product is currently in phase 1.
- Undergraduate biomedical engineering students at Johns Hopkins University have created a new drug delivery system based on the thin-film technology used by a breath freshener. Laced with a vaccine against rotavirus, the strips could be used to provide the vaccine to infants in impoverished areas.

Advantages

The design of thin film as an oral drug delivery technology offers several advantages over other modes of drug delivery, such as ingestible tablets, chewable tablets, orally dissolving tablets, softgels, liquids or inhalants:

- The sublingual and buccal delivery of a drug via thin disolvable film has the potential to improve the onset of action, lower the dosing, and enhance the efficacy and safety profile of the medicament.
  - All tablet dosage forms, softgels and liquid formulations primarily enter the blood stream via the gastrointestinal tract, which subjects the drug to degradation from stomach acid, bile, digestive enzymes and other first-pass effects. As a result, such formulations often require higher doses and generally have a delayed onset of action.
  - Conversely, buccal and sublingual thin-film drug delivery can avoid these issues and yield quicker onsets of action at lower doses.
- Thin film is more stable, durable and quicker dissolving than other conventional dosage forms.
- Thin film enables improved dosing accuracy relative to liquid formulations since every strip is manufactured to contain a precise amount of the drug.

- Thin film not only ensures more accurate administration of drugs but also can improve compliance due to the intuitive nature of the dosage form and its inherent ease of administration. These properties are especially beneficial for pediatric, geriatric and neurodegenerative disease patients where proper and complete dosing can be difficult.
- Thin film's ability to dissolve rapidly without the need for water provides an alternative to patients with swallowing disorders and to patients suffering from nausea, such as those patients receiving chemotherapy.
- Thin film drug delivery has the potential to allow the development of sensitive drug targets that may otherwise not be possible in tablet or liquid formulations.
- From a commercial perspective thin film drug delivery technology offers an opportunity to extend revenue lifecycles for pharmaceutical companies whose drug patent is expiring and will soon be vulnerable to generic competition.
- Pharmaceutical drugs can be contained within an abuse-deterrent film matrix that cannot be crushed or injected by patients, and rapidly absorbs under the tongue to ensure compliance.
- Facilitates absorption 3 to 10 times greater than an oral tablet—only surpassed by hypodermic injection.

Self-Microemulsifying Drug Delivery System

A self-microemulsifying drug delivery system (SMEDDS) is a drug delivery system that uses a microemulsion achieved by chemical rather than mechanical means. That is, by an intrinsic property of the drug formulation, rather than by special mixing and handling. It employs the familiar ouzo effect displayed by anethole in many anise-flavoured liquors. Microemulsions have significant potential for use in drug delivery, and SMEDDS (including so-called "U-type" microemulsions) are the best of these systems identified to date. SMEDDS are of particular value in increasing the absorption of lipophilic drugs taken by mouth.

Microemulsions are clear, thermodynamically stable, isotropic liquid mixtures of oil, water and surfactant, frequently in combination with a cosurfactant. The aqueous phase may contain salt(s) and/or other ingredients, and the "oil" may actually be a complex mixture of different hydrocarbons and olefins. In contrast to ordinary emulsions, microemulsions form upon simple mixing of the components and do not require the high shear conditions generally used in the formation of ordinary emulsions. The three basic types of microemulsions are direct (oil dispersed in water, o/w), reversed (water dispersed in oil, w/o) and bicontinuous.

In ternary systems such as microemulsions, where two immiscible phases (water and 'oil') are present with a surfactant, the surfactant molecules may form a monolayer at the interface between the oil and water, with the hydrophobic tails of the surfactant molecules dissolved in the oil phase and the hydrophilic head groups in the aqueous phase.

SMEDDS in research or development include formulations of the drugs anethole trithione, oridonin, curcumin, vinpocetine, tacrolimus, berberine hydrochloride, nobiletin, piroxicam, anti-malaria drugs beta-Artemether and halofantrine, anti-HIV drug UC 781, nimodipine, exemestane, anti-cancer drugs 9-nitrocamptothecin (9-NC) paclitaxel, and seocalcitol, alprostadil (intraurethral use), probucol, itraconazole, fenofibrate, acyclovir, simvastatin, xibornol, silymarin, alpha-asarone, enilconazole, puerarin (an isoflavone found in *Pueraria lobata*), atorvastatin, heparin, carvedilol, ketoconazole, gentamicin, labrasol, flurbiprofen, celecoxib, danazol, cyclosporine, and idebenone.

Actual applications of *Self-microemulsifying drug delivery system'* (SMEDDS) remain rare. The first drug marketed as a SMEDDS was cyclosporin, and it had significantly improved bioavailability compared with the conventional solution. In the last decade, several SMEDDS loaded with antiviral drugs (ritonavir, saquinavir) were tested for treatment of HIV infection, but the relative improvement in clinical benefit was not significant. The SMEDDS formulation of ritonavir (soft capsules) has been withdrawn in some countries.

SMEDDS offer numerous advantages: spontaneous formation, ease of manufacture, thermodynamic stability, and improved solubilization of bioactive SMEDDS. Improved solubility contributes to faster release rates and greater bioavailability. For many drugs taken by mouth, faster release rates improve the drug acceptance by consumers. Greater bioavailability means that less drug need be used; this may lower cost, and does lower the stomach irritation and toxicity of drugs taken by mouth.

For oral use, SMEDDS may be formulated as liquids or solids, the solids packaged in capsules or tablets. Limited studies comparing these report that in terms of bioavailability liquid SMEDDS are superior to solid SMEDDS, which are superior to conventional tablets. Liquid SMEDDS have also shown value in injectable (IV and urethral) formulations and in a topical (oral) spray.

Acoustic Targeted Drug Delivery

Acoustic targeted drug delivery (ATDD) is a method that uses ultrasound energy to enhance the transport of molecules into and/or across specific tissues. Generally this class of ultrasound energy falls under the class of therapeutic ultrasound, and ranges in ultrasonic frequencies of 1-20 MHz and Sound intensitys of 0-30 watts/cm The use of ATDD in conjunction with local drug delivery by injection, topical application and convection enhanced delivery shows promise to significantly enhance the treatment of various diseases in the human body by specifically targeting the drug into the tissue.

How ATDD Works

Focused ultrasound similar to High intensity focused ultrasound, but at much lower powers is applied to tissues in conjunction with the local application of pharmological agent.

The ultrasound source (or transducer) is pulsed on and off, and moved in a defined pattern to *sonicate* the tissue of interest. The pulsing of the ultrasound serves two purposes 1. to agitate the tissue matrix by extending and compressing it, and 2. to keep the tissue from absorbing too much energy and ablating it (or causing necrosis).

The manual movement of the transducer allows for control of the *sonicated volume* of tissue and where the drug may penetrate more readily.

ATDD may be applied:

- Geometrically, for example with a lens or with a spherically curved transducer.
- Electronically, by adjusting the relative phases of elements in an array of transducers (a "phased array"). By dynamically adjusting the electronic signals to the elements of a phased array, the beam can be steered to different locations, and aberrations due to tissue structures can be corrected.

ATDD and Topical Drug Delivery

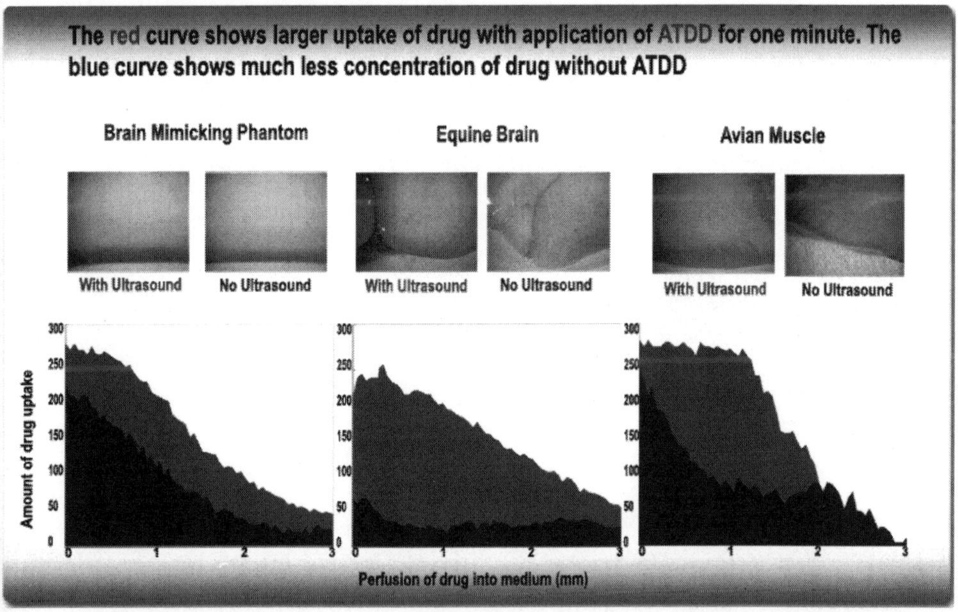

Neural Drug Delivery Systems

Neural drug delivery is the next step beyond the basic addition of growth factors to nerve guidance conduits. Drug delivery systems allow the rate of growth factor release to be regulated over time, which is critical for creating an environment more closely representative of in vivo development environments.

Delivery systems can have many different uses; however, the emphasis of the discussion here will be on the need for drug delivery systems in nervous system injuries.

First Principles

A major challenge of delivering growth factors to lesions in the spinal cord is the high clearance and removal due to the high turnover rate of the cerebrospinal fluid.

One way to overcome this is to use a drug delivery system that slowly releases the growth factors over a prolonged period of time. One previously developed affinity-based drug delivery system regulates the slow release of growth factors by binding them to heparin in fibrin gels. In addition to the fibrin gel there are three main components to the heparin-based delivery system (HBDS): a synthetic linker peptide, the polysulfated glycosaminoglycan heparin, and the growth factors being delivered.

The synthetic linker peptide is covalently crosslinked to the fibrin gels and is noncovalently linked to heparin, which noncovalently links the desired growth factor to be delivered.

The release of the growth factor from heparin is mediated by cell-activated plasmin degradation. When axons extend into the gel, they activate the conversion of plasminogen to plasmin, thereby initiating plasmin degradation of the fibrin gel and release of a growth factor such as NT-3.

This drug delivery system has been used extensively to deliver growth factors with a high affinity for heparin, such as basic fibroblast growth factor. It has even been used for some growth factors with lower affinities such as nerve growth factor and NT-3. When HBDS was used to deliver NGF and NT-3, dorsal root ganglion neurite extension was enhanced by 75% and 54% over the controls which included the growth factors freely in media. This drug delivery system presents promisingpotential for in vivo use as an effective way to deliver growth factors to spinal cord lesions and thus promote regeneration.

Delivering Growth Factors

• It is also challenging to deliver growth factors to the central nervous system, because most proteins are not able to cross the blood–brain barrier and therefore, cannot be delivery via systemic administration. The most desirable method of delivering growth factors such as nerve growth factor is locally because then only nanogram quantities are needed to supply therapeutic levels to the targeted cell population.

A controlled delivery system is needed in order to supply growth factor locally. Another drug delivery system that has been well characterized is poly(ethylene-co-vinyl acetate) (EVAc). EVAc matrices have been approved for use the Food and Drug Administration.

Protein release occurs mainly by diffusion, which can be regulated by changing the mass within the polymer matrix.

In order to better determine appropriate dosages from EVAc matrices, a study was done to determine how cell density and cell type affect the rate at which a delivered protein is eliminated; NGF was released into PC12 and developing fetal brain cells distributed throughout collagen gels. It was found that both the cell type and the cell density affected the rate of NGF elimination from the gels, suggesting that elimination is a cell-mediated process.

The results were compared to a model that was developed to predict the effects of cell type and density on elimination. One specific reason for differences between the data and the model predictions could be because the model does not account for the change increases in cell number due to replication. This study is important for enhancing the ability to accurate deliver the desired dosage to the target tissue based on the amount of growth factor initially in the delivery matrices.

Drug Carrier

Drug carriers are substances that serve as mechanisms to improve the delivery and the effectiveness of drugs. Drug carriers are used in sundry drug delivery systems such as:

- controlled-release technology to prolong in vivo drug actions;
- decrease drug metabolism, and
- reduce drug toxicity.

Carriers are also used in designs to increase the effectiveness of drug delivery to the target sites of pharmacological actions.

Nanocarriers

A nanocarrier is nanomaterial being used as a transport module for another substance, such as a drug. Commonly used nanocarriers include micelles, polymers, carbon-based materials, liposomes and other substances. Nanocarriers are currently being studied for their use in drug delivery and their unique characteristics demonstrate potential use in chemotherapy. Nanocarriers range from sizes of diameter 1–100 nm. Because of their small size, nanocarriers can deliver drugs to otherwise inaccessible sites around the body. Since nanocarriers are so small, it is oftentimes difficult to provide large drug doses using them.

The emulsion techniques used to make nanocarriers also often result in low drug loading and drug encapsulation, providing a difficulty for the clinical use.

Types

Nanocarriers discovered thus far include polymer conjugates, polymeric nanoparticles, lipid-based carriers, dendrimers, carbon nanotubes, and gold nanoparticles. Lipid-based carriers include both liposomes and micelles. Examples of gold nanoparticles are gold nanoshells and nanocages. Different types of nanomaterial being used in nanocarriers allows for hydrophobic and hydrophilic drugs to be delivered throughout the body.

Since the human body contains mostly water, the ability to deliver hydrophobic drugs effectively in humans is a major therapeutic benefit of nanocarriers. Micelles are able to contain either hydrophilic or hydrophobic drugs depending on the orientation of the phospholipid molecules. Some nanocarriers contain nanotube arrays allowing them to contain both hydrophobic and hydrophilic drugs.

One potential problem with nanocarriers is unwanted toxicity from the type of nanomaterial being used. Inorganic nanomaterial can also be toxic to the human body if it accumulates in certain cell organelles. New research is being conducted to invent more effective, safer nanocarriers. Protein based nanocarriers show promise for use therapeutically since they occur naturally, and generally demonstrate less cytotoxicity than synthetic molecules.

Targeted Drug Delivery

Nanocarriers are useful in the drug delivery process because they can deliver drugs to site-specific targets, allowing drugs to be delivered in certain organs or cells but not in others. Site-specificity is a major therapeutic benefit since it prevents drugs from being delivered to the wrong places. Nanocarriers show promise for use in chemotherapy because they can help decrease the adverse, broader-scale toxicity of chemotherapy on healthy, fast growing cells around the body. Since chemotherapy drugs can be extremely toxic to human cells, it is important that they are delivered to the tumor without being released into other parts of the body. Four methods in which nanocarriers can deliver drugs include passive targeting, active targeting, pH specificity, and temperature specificity.

Passive Targeting

Passive targeting refers to a nanocarrier's ability to travel down a tumor's vascular system, become trapped, and accumulate in the tumor. This accumulation is caused by the enhanced permeability and retention effect which refers to the poly(ethylene oxide) (PEO) coating on the outside of many nanocarriers. PEO allows nanocarriers to travel through the leaky vasculature of a tumor, where they are unable to escape.

The leaky vasculature of a tumor is the network of blood vessels that form in a tumor, which contain many small pores. These pores allow nanocarriers in, but also contain many bends that allow the nanocarriers to become trapped. As more nanocarriers become trapped, the drug accumulates at the tumor site. This accumulation cause large doses of the drug to be delivered directly to the tumor site. PEO may also have some adverse affects on cell-nanocarrier interactions, weakening the effects of the drug, since many nanocarriers must be incorporated into the cells before the drugs can be released.

Active Targeting

Active targeting involves the incorporation of targeting modules such as ligands or antibodies on the surface of nanocarriers that are specific to certain types of cells around the body.

Nanocarriers have such a high surface-area to volume ratio allowing for multiple ligands to be incorporated on their surfaces. These targeting modules allow for the nanocarriers to be incorporated directly inside of cells, but also have some drawbacks.

Ligands may cause nanocarriers to become slightly more toxic due to non-specific binding, and positive charges on ligands may decrease drug delivery efficiency once inside of cells. Active targeting has been shown to help overcome multi-drug resistance in tumor cells.

pH Specificity

Certain nanocarriers will only release the drugs they contain in specific pH ranges. pH specificity also allows nanocarriers to deliver drugs directly to a tumor site. Tumors are generally more acidic than normal human cells, with a pH around 6.8.

Normal tissue has a pH of around 7.4. Nanocarriers that only release drugs at certain pH ranges can therefore be used to release the drug only within acidic tumor environments. High acidic environments cause the drug to be released due to the acidic environment degrading the structure of the nanocarrier.

These nanocarriers will not release drugs in neutral or basic environments, effectively targeting the acidic environments of tumors while leaving normal body cells untouched. This pH sensitivity can also be induced in micelle systems by adding copolymer chains to micelles that have been determined to act in a pH independent manor.

These micelle-polymer complexes also help to prevent cancer cells from developing multi-drug resistance. The low pH environment triggers a quick release of the micelle polymers, causing a majority of the drug to be released at once, rather than gradually like other drug treatments.

This quick release mechanism significantly decreases the time it takes for anticancer drugs to kill a tumor, effectively preventing the tumor from having time to undergo mutations that would render it drug resistant.

Temperature Specificity

Some nanocarriers have also been shown to deliver drugs more effectively at certain temperatures. Since tumor temperatures are generally higher than temperatures throughout the rest of the body, around 40 °C, this temperature gradient helps act as safeguard for tumor-specific site delivery.

Uses

Most of research on nanocarriers is being applied to their potential use in drug delivery, especially in chemotherapy. Since nanocarriers can be used to specifically target the small pores, lower pH's, and higher temperatures of tumors, they have the potential to lower the toxicity of many chemotherapy drugs. Also, since almost 75% of anticancer drugs are hydrophobic, and therefore demonstrate difficulty in delivery inside human cells, the use of micelles to stabilize, and effectively mask the hydrophobic nature of hydrophobic drugs provides new possibilities for hydrophobic anticancer drugs.

Bovine Submaxillary Mucin Coatings

Bovine submaxillary mucin (BSM) coatings are a surface treatment provided to biomaterials intended to reduce the growth of disadvantageous bacteria such as S. epidermidis, E. coli, and Candida albicans. BSM is a substance extracted from the fresh salivary glands of cows. It exhibits unique physical properties, such as high molecular weight and amphiphilicity, that allow it to be used for many biomedical applications.

Each species possesses mucin-secreting submaxillary glands. Currently, eight different mucins have been identified for humans. However, it is the mucin from bovine and porcine sources that have been used in several biomaterial applications.

The most common use of BSM is in coatings for implanted materials. In such applications, the adsorption characteristics of BSM are integral to the behaviour of materials in vivo. Survival and rejection of an implant are highly dependent on surface modifications that dictate the interfacial interaction between a material and the body. Thus, BSM adsorption increases biocompatibility.

Epithelial tissue lines surfaces throughout the body. A mucous gel layer composed of protein coats this tissue. The main function of the layer is protection from mechanical stress, dehydration, and bacterial infections. It also plays a role in cellular signaling, cellular interactions, and pH regulation. The two main components of the gel are mucins and water, with the mucins forming the backbone of the mucous.

Mucin originates from the mucus secretions from submaxillary glands, which are salivary glands that are located under the floor of the mouth. The secreted mucin assists in digestion by coating the bolus such that it travels easily through the digestive tract.

Mucins are constituents of one class of glycoproteins: sialic acid-containing glycoproteins, or mucoproteins. They have a high molecular weight and exist as either membrane-bound or secretory mucins.

The membrane-bound type have a hydrophilic area that extends along a membrane and attaches to cellular surfaces. Secretory mucins are the main components of the mucous gel layer that coats the epithelium.

Physical Characteristics

While mucous is mostly composed of water, its structural and rheological properties are characterized by mucin. Under an atomic force microscope, BSM appears as a composition of dumbbell shaped fibers at heights of approximately 1 nm. BSM can be purchased as a powder, in which case it appears white to light brown.

Mechanical Properties

Mucin is a fibrous matrix with gel-like properties. When mixed into an aqueous solution, higher concentrations of mucin result in lower surface tension values. BSM has a hydration enthalpy of -20 kJ/mol and an RH level for glass transition of 60-70%.

Table: *Properties of Mucin from Various Species*

Source of Mucin	Molecular Weight [kDa]	Γ [mg/m^2] after 24h adsorption to hydrophilic silica with 5 mg/ml mucin	Γ [mg/m^2] after 24h adsorption to hydrophobic silica with 5 mg/ml mucin
Bovine Submaxillary	4000	0.25	2
Porcine Submaxillary	500-250	1.15	2
Human Sublingual (High Molecular Weight)	>1000	-	5

Table: *Surface Properties of BSM Coated versus Non-Coated Materials*

	θ, Contact Angle of Water on PMMA	γ, Surface Tension on PBS
Non-Coated	69.4±0.2°	70.2 [mN/m]
Coated	50.2±4.0°	46.5-47.3 [mN/m]

Table: *Surface Properties of Various Solutions of Mucin and NaCl*

	λ, Debye Length [nm]	***κ, Adsorption [nm]***	***δ, Surface Density of Total Adsorbed Mucin [mg/m²***
Mucin 0.1 g/L + NaCl 10^{-3} M	41	0.024	2.4
Mucin 0.2 g/L + NaCl 10^{-3} M	43	0.105	4.3
Mucin 0.1 g/L + NaCl 0.15 M	41	0.024	4.5
Mucin 0.05 g/L + NaCl 0.15 M	110	0.009	2.3

Chemical Composition

Being large glycoproteins, mucins have high carbohydrate content, contributing to their fibrous structure.

These carbohydrates branch off of polypeptide chains in the form of oligosaccharides including N-acetylgalactosamine, N-acetylglucosamine, fucose, galactose, and sialic acid. The serine and threonine hydroxyl groups link to the polypeptide chains via O-glycosidic linkages.

Interaction with Surfaces

Mucins interact with surfaces in a unique way due to the amphiphilic properties the molecules exhibit.

Adsorption of BSM to Solid Surfaces and Types of Bonding

Mucins adhere well to polymeric surfaces through various types of bonding:

(1) covalent bond attachment

(2) electrostatic interaction, which requires matching of charge groups between the polymer and mucus;

(3) hydrogen bonding; and

(4) hydrophobic interactions .

These interactions require very close contact between the surface and the mucus in order to facilitate strong bonding.

Most mucins adsorb to solid surfaces fairly easily without surface preparation due to the attractive interaction of the surface with the non-glycosylated region of the molecule. Surface force measurements reveal that long range steric repulsion forces exist between mucin coated surfaces and can be detected at a distance of up 100 nm or more. This indicates that mucin segments extend into the ambient solution at a long distance. Because mucin is amphiphilic, the non-glycosylated regions of the molecule interacts with the surface and the glycosylated region interacts with the ambient solution causing the steric repulsion. The repulsive forces applied by the mucin creates anti-adhesive properties which can suppress cell adhesion to surfaces.

The efficiency of mucin adherence to hydrophobic surfaces has been analyzed by applying a coating of mucin and measuring the number of mucins remaining on the hydrophobic material after rinsing. The amount of mucin removed was a small fraction of the remaining material.

A different method to determine the amount of mucin adsorption to a solid surface involves determining the refractive index of the clean solid surface and comparing that number to the refractive index with adsorbed mucin on the surface. The adsorbtion of mucin will alter the refractive index which can be used to calculate the amount of mucin molecules in the adsorbed layer.

Equation to estimate the Index of Refraction:

$$n = n_s + dn/dc.c$$

n= Refractive Index

n_s = Refractive Index of Solvent

dn/dc= Refractive Index Increment

c= Concentrations of Dissolved Molecules

Equation to estimate the amount of molecules in the adsorbed layer:

$$\Gamma = n - n_s/dn/dc.d$$

Γ= Surface Excess

n= Refractive Index

n_s = Refractive Index of Solvent

dn/dc= Refractive Index Increment

d= Ellipsometric Thickness

Hydrophobic and Electrostatic Interactions

Proteins usually readily adsorb to solid surfaces from aqueous solutions to form a protein film, typically mixed with water. Many types of interactions can mediate the adsorption; among these, hydrophobic and electrostatic interactions have been identified as central factors determining protein adsorption to surfaces and the repulsive reaction to molecules in solution. In addition, structural rearrangements of proteins as well as hydrogen bonding are suggested to influence the adsorption.

The glycosylated region of mucins extend into the water which forces the molecules to extend into the ambient solution in a random coil conformation. Additionally, the high molecular weight of mucin allows the molecules to overlap and entangle when present in low concentrations.

PH Dependence of Adsorption

Empirical data has shown that pH has an influence on the adsorption of mucin to surfaces. At pH 4, the adsorbed amount of mucin was higher, but the thickness of the layer was lower when compared to pH 6 environments. The increased amount of mucin adsorption can be explained by the high amount of protonated amino groups that are present at the pH 4 level that form electrostatic bonds with negatively charged groups in the molecule.

At lower pH values and low ionic strength, the adsorbed amount of BSM to a surface increases.

The lower thickness level seen a lower pH values can be explained by a stronger attractive interaction of mucin to the surface causing the particles to deform more readily to increase amount of adsorption.

Biocompatibility

Because BSM is frequently applied as a coating on a material that will interact with the human body, it is important to minimize disadvantageous interactions with tissues and cells. Two considerations include reducing tissue reactivity, as in inflammatory and immune responses, and preventing the adsorption of particles, like bacteria.

Tissue Reactivity

BSM-coated polyurethane samples were implanted in sheep in order to assess if the coatings caused minimal host responses during a 30 day implantation. After inspection of the explanted samples, it was revealed that the host tissue had lower inflammatory responses and minimal capsule formation with the BSM-coated polyurethane samples than the uncoated samples.

These results make the BSM coatings appealing for use in biomedical applications where synthetic materials directly interface with tissues.

Mucins are also considered useful in tissue related applications because they are natural and biological, can form hydrogels, show resistance to proteolytic degradation, and have good adhesion to surfaces while repelling molecules at the ambient solution interface.

Also, the possibility to allow the uptake of certain proteins makes the BSM coatings very appealing for biomedical uses because correct protein uptake can potentially increase the recovery speed in patients.

Prevention of Adsorbing Particles

In the previous sections, it has been described that mucin is useful in anti-adhesive and repelling applications. This is because BSM adheres well to surfaces, but the non polar ends of the molecules that extend into the ambient solution exhibit a repulsive behaviour that prevents the external particles from reaching the surface for adsorption.

In the activity of preventing more adsorbed particles, the BSM layer only has to be 0.3 mg/m^2 thick to completely prevent adsorption.

Applications

Implant Coatings

BSM is an active area of research for several biomedical engineering applications. Due to its adsorption on solid surfaces, it is useful in vivo to prevent bacterial infections. BSM and bacteria compete for adsorption sites; mucin adhesion blocks bacterial adhesion on surfaces. Consequently, the risk of bacterial attachment and infection is reduced.

While bacterial infections are common concerns in the medical, dental, and food industries, they are the primary cause of device failures in biological implants. To combat this issue, BSM has been used as a coating for a poly(acrylic acid-b-methyl methacrylate) (PAA-b-PMMA) diblock copolymer.

The copolymer provides two mechanisms that contribute to the bacterial resistance of the biomaterial. First, the PAA block assists the adsorption of BSM to the surface. Second, the PMMA block prevents the dissolution of BSM in aqueous environments. This behaviour is imperative in the lubrication of cardiovascular stents and urinary catheters as they are inserted within patients. Due to the adsorption of BSM, the hydrophobicity of the surface decreases as well as the incidence of bacterial adhesion.

Drug Delivery Systems

Drug delivery is another application that utilises BSM coatings. Drug delivery systems employ pharmaceutical products within encapsulated compartments.

These compartments are composed of hydrophilic polymers that enable the release of a particular drug at a specified rate and release profile. BSM coatings are advantageous in mucoadhesive films, which release drugs to surrounding mucosal tissue after implantation. Such delivery systems have been made specifically with PAA polymers and BSM coatings.

Just like the application above, the PAA polymer improved the adsorption of BSM on its surface to prevent bacterial adhesion and potential infection. Mucin interaction with the pharmaceutical product itself and other proteins encountered in the body is vital for the drug delivery process. The behaviour of BSM and drugs is investigated to ensure that one element does not disrupt the function of the other. Coatings are also synthesized, tested, and analyzed such that various proteins in the body do not adversely affect BSM adsorption and bacterial resistance.

Hydrogels

BSM has also been used for the fabrication of hydrogels. Hydrogels are crosslinked hydrophilic polymer matrices in water, which is the dispersion medium. The properties of BSM are ideal for hydrogel formation. Its glycosylated regions interact with water, forming elongated random coils.

Additionally, its high molecular weight enables it to overlap at low concentrations. These synergistic behaviours create effective gel matrices that are suitable for several biomedical applications, such as scaffolds, medical electrodes, and drug delivery systems.

7

Nanotechnology and Tissue Engineering

Tissue engineering is the use of a combination of cells, engineering and materials methods, and suitable biochemical and physio-chemical factors to improve or replace biological functions. While it was once categorized as a sub-field of bio materials, having grown in scope and importance it can be considered as a field in its own right.

While most definitions of tissue engineering cover a broad range of applications, in practice the term is closely associated with applications that repair or replace portions of or whole tissues (i.e., bone, cartilage, blood vessels, bladder, skin, muscle etc.). Often, the tissues involved require certain mechanical and structural properties for proper functioning. The term has also been applied to efforts to perform specific biochemical functions using cells within an artificially-created support system (e.g. an artificial pancreas, or a bio artificial liver). The term *regenerative medicine* is often used synonymously with tissue engineering, although those involved in regenerative medicine place more emphasis on the use of stem cells or progenitor cells to produce tissues.

A commonly applied definition of tissue engineering, as stated by Langer and Vacanti, is "an interdisciplinary field that applies the principles of engineering and life sciences toward the development of biological substitutes that restore, maintain, or improve tissue function or a whole organ". Tissue engineering has also been defined as "understanding the principles of tissue growth, and applying this to produce functional replacement tissue for clinical use." A further description goes on to say that an "underlying supposition of tissue engineering is that the employment of natural biology of the system will allow for greater success in developing therapeutic strategies aimed at the replacement, repair, maintenance, and/or enhancement of tissue function."

Powerful developments in the multidisciplinary field of tissue engineering have yielded a novel set of tissue replacement parts and implementation strategies. Scientific advances in biomaterials, stem cells, growth and differentiation factors, and biomimetic environments have created unique opportunities to fabricate tissues in the laboratory from combinations of engineered extracellular matrices ("scaffolds"), cells, and biologically active molecules. Among the major challenges now facing tissue engineering is the need for more complex functionality, as well as both functional and biomechanical stability in laboratory-grown tissues destined for transplantation. The continued success of tissue engineering, and the eventual development of true human replacement parts, will grow from the convergence of engineering and basic research advances in tissue, matrix, growth factor, stem cell, and developmental biology, as well as materials science and bio informatics.

In 2003, the NSF published a report entitled "The Emergence of Tissue Engineering as a Research Field", which gives a thorough description of the history of this field.

Examples:

- Bioartificial windpipe : The first procedure of regenerative medicine of an implantation of a "bioartificial" organ.
- In vitro meat: Edible artificial animal muscle tissue cultured *in vitro.*
- Bioartificial liver device: several research efforts have produced hepatic assist devices utilising living hepatocytes.
- Artificial pancreas: research involves using islet cells to produce and regulate insulin, particularly in cases of diabetes.
- Artificial bladders: Anthony Atala (Wake Forest University) has successfully implanted artificially grown bladders into seven out of approximately 20 human test subjects as part of a long-term experiment.
- Cartilage: lab-grown tissue was successfully used to repair knee cartilage.
- Scaffold-free cartilage: Cartilage generated without the use of exogenous scaffold material. In this methodology, all material in the construct is cellular or material produced directly by the cells themselves.
- Doris Taylor's heart in a jar
- Tissue-engineered airway
- Tissue-engineered vessels
- Artificial skin constructed from human skin cells embedded in collagen
- Artificial bone marrow

- Artificial bone
- Artificial penis
- Oral mucosa tissue engineering
- Foreskin

Cells as Building Blocks

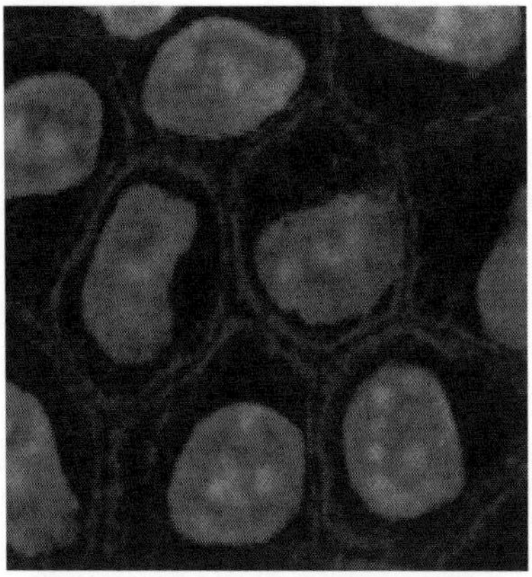

Figure: *Stained cells in culture*

Tissue engineering utilises living cells as engineering materials. Examples include using living fibroblasts in skin replacement or repair, cartilage repaired with living chondrocytes, or other types of cells used in other ways. Cells became available as engineering materials when scientists at Geron Corp. discovered how to extend telomeres in 1998, producing immortalized cell lines. Before this, laboratory cultures of healthy, noncancerous mammalian cells would only divide a fixed number of times, up to the Hayflick limit.

Extraction

From fluid tissues such as blood, cells are extracted by bulk methods, usually centrifugation or apheresis. From solid tissues, extraction is more difficult. Usually the tissue is minced, and then digested with the enzymes trypsin or collagenase to remove the extracellular matrix that holds the cells. After that, the cells are free floating, and extracted using centrifugation or apheresis. Digestion with trypsin is very dependent on temperature. Higher temperatures digest the matrix faster, but create more damage. Collagenase is less temperature dependent, and damages fewer cells, but takes longer and is a more expensive reagent.

Types of Cells

Cells are often categorized by their source:

- *Autologous* cells are obtained from the same individual to which they will be reimplanted. Autologous cells have the fewest problems with rejection and pathogen transmission, however in some cases might not be available. For example in genetic disease suitable autologous cells are not available. Also very ill or elderly persons, as well as patients suffering from severe burns, may not have sufficient quantities of autologous cells to establish useful cell lines. Moreover since this category of cells needs to be harvested from the patient, there are also some concerns related to the necessity of performing such surgical operations that might lead to donor site infection or chronic pain. Autologous cells also must be cultured from samples before they can be used: this takes time, so autologous solutions may not be very quick. Recently there has been a trend towards the use of mesenchymal stem cells from bone marrow and fat. These cells can differentiate into a variety of tissue types, including bone, cartilage, fat, and nerve. A large number of cells can be easily and quickly isolated from fat, thus opening the potential for large numbers of cells to be quickly and easily obtained.

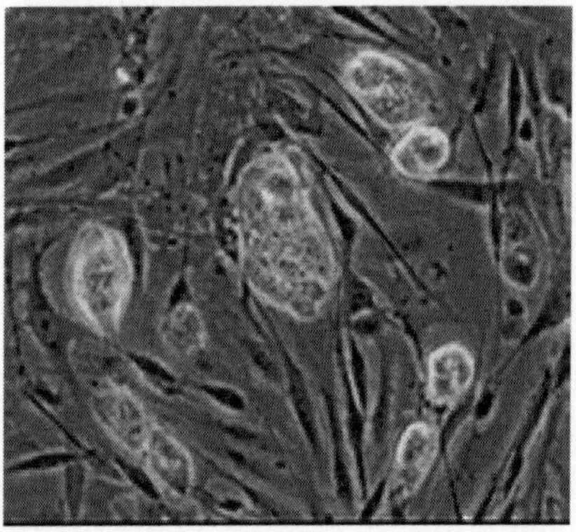

Figure: *Mouse embryonic stem cells*

- *Allogeneic* cells come from the body of a donor of the same species. While there are some ethical constraints to the use of human cells for *in vitro* studies, the employment of dermal fibroblasts from human foreskin has been demonstrated to be immunologically safe and thus a viable choice for tissue engineering of skin.
- *Xenogenic* cells are these isolated from individuals of another species. In particular animal cells have been used quite extensively in experiments aimed at the construction of cardiovascular implants.
- *Syngenic* or *isogenic* cells are isolated from genetically identical organisms, such as twins, clones, or highly inbred research animal models.
- *Primary* cells are from an organism.
- *Secondary* cells are from a cell bank.
- *Stem cells* are undifferentiated cells with the ability to divide in culture and give rise to different forms of specialised cells. According to their

source stem cells are divided into "adult" and "embryonic" stem cells, the first class being multipotent and the latter mostly pluripotent; some cells are totipotent, in the earliest stages of the embryo. While there is still a large ethical debate related with the use of embryonic stem cells, it is thought that another alternative source - induced stem cells may be useful for the repair of diseased or damaged tissues, or may be used to grow new organs.

Scaffolds

Cells are often implanted or 'seeded' into an artificial structure capable of supporting three-dimensional tissue formation. These structures, typically called scaffolds, are often critical, both *ex vivo* as well as *in vivo*, to recapitulating the *in vivo* milieu and allowing cells to influence their own microenvironments. Scaffolds usually serve at least one of the following purposes:

- Allow cell attachment and migration
- Deliver and retain cells and biochemical factors
- Enable diffusion of vital cell nutrients and expressed products
- Exert certain mechanical and biological influences to modify the behaviour of the cell phase

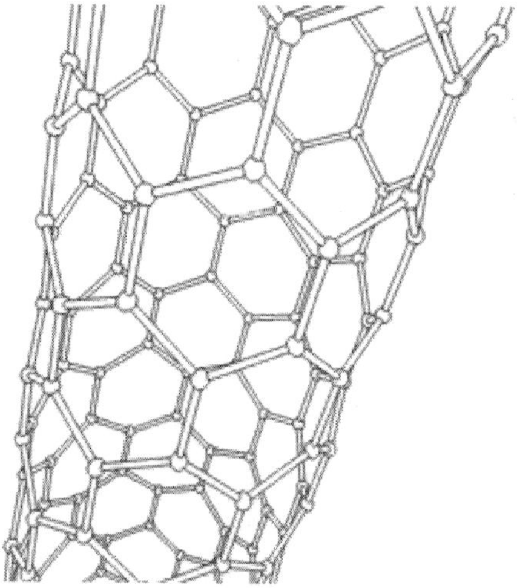

Figure: *This animation of a rotating Carbon nanotube shows its 3D structure. Carbon nanotubes are among the numerous candidates for tissue engineering scaffolds since they are biocompatible, resistant to biodegradation and can be functionalized with biomolecules. However, the possibility of toxicity with non-biodegradable nano-materials is not fully understood.*

To achieve the goal of tissue reconstruction, scaffolds must meet some specific requirements. A high porosity and an adequate pore size are necessary to facilitate cell seeding and diffusion throughout the whole structure of both cells and nutrients. Biodegradability is often an essential factor since scaffolds should preferably be absorbed by the surrounding tissues without the necessity of a surgical removal. The rate at which degradation occurs has to coincide as much as possible with the rate of tissue formation: this means that while cells are fabricating their own natural matrix structure around themselves, the scaffold is able to provide structural integrity within the body and eventually it will break down leaving the neotissue, newly formed tissue which will take over the mechanical load. Injectability is also important for clinical uses. Recent research on organ printing is showing how crucial a good control of the 3D environment is to ensure reproducibility of experiments and offer better results.

Materials

Many different materials (natural and synthetic, biodegradable and permanent) have been investigated. Most of these materials have been known in the medical field before the advent of tissue engineering as a research topic, being already employed as bioresorbable sutures. Examples of these materials are collagen and some polyesters.

New biomaterials have been engineered to have ideal properties and functional customization: injectability, synthetic manufacture, biocompatibility, non-immunogenicity, transparency, nano-scale fibers, low concentration, resorption rates, etc. PuraMatrix, originating from the MIT labs of Zhang, Rich, Grodzinsky and Langer is one of these new biomimetic scaffold families which has now been commercialized and is impacting clinical tissue engineering.

A commonly used synthetic material is PLA - polylactic acid. This is a polyester which degrades within the human body to form lactic acid, a naturally occurring chemical which is easily removed from the body. Similar materials are polyglycolic acid (PGA) and polycaprolactone (PCL): their degradation mechanism is similar to that of PLA, but they exhibit respectively a faster and a slower rate of degradation compared to PLA.

Scaffolds may also be constructed from natural materials: in particular different derivatives of the extracellular matrix have been studied to evaluate their ability to support cell growth. Proteic materials, such as collagen or fibrin, and polysaccharidic materials, like chitosan or glycosaminoglycans (GAGs), have all proved suitable in terms of cell compatibility, but some issues with potential immunogenicity still remains. Among GAGs hyaluronic acid, possibly in combination with cross linking agents (e.g. glutaraldehyde, water soluble carbodiimide, etc...), is one of the possible choices as scaffold material. Functionalized groups of scaffolds may be useful in the delivery of small molecules (drugs) to specific tissues. Another form of scaffold under investigation is decellularised tissue extracts whereby the remaining cellular remnants/extracellular matrices act as the scaffold.

A 2009 study by Ratmir et al. aimed to improve in vivo-like conditions for 3D tissue via "stacking and de-stacking layers of paper impregnated with suspensions of cells in extracellular matrix hydrogel, making it possible to control oxygen and nutrient gradients in 3D, and to analyze molecular and genetic responses". It is possible to manipulate gradients of soluble molecules, and to characterize cells in these complex gradients more effectively than conventional 3D cultures based on hydrogels, cell spheroids, or 3D perfusion reactors. Different thicknesses of paper and types of medium can support a variety of experimental environments. Upon deconstruction, these sheets can be useful in cell-based high-throughput screening and drug discovery.

Synthesis

A number of different methods have been described in literature for preparing porous structures to be employed as tissue engineering scaffolds. Each of these techniques presents its own advantages, but none are free of drawbacks.

Nanofiber Self-Assembly: Molecular self-assembly is one of the few methods for creating biomaterials with properties similar in scale and chemistry to that of the natural in vivo extracellular matrix (ECM). Moreover, these hydrogel scaffolds have shown superiority in vivo toxicology and biocompatibility compared to traditional macroscaffolds and animal-derived materials.

Textile Technologies: These techniques include all the approaches that have been successfully employed for the preparation of non-woven meshes of different polymers. In particular, non-woven polyglycolide structures have been tested for tissue engineering applications: such fibrous structures have been found useful to grow different types of cells. The principal drawbacks are related to the difficulties in obtaining high porosity and regular pore size.

Solvent Casting & Particulate Leaching (SCPL): This approach allows for the preparation of porous structures with regular porosity, but with a limited thickness. First, the polymer is dissolved into a suitable organic solvent (e.g. polylactic acid could be dissolved into dichloromethane), then the solution is cast into a mold filled with porogen particles. Such porogen can be an inorganic salt like sodium chloride, crystals of saccharose, gelatin spheres or paraffin spheres. The size of the porogen particles will affect the size of the scaffold pores, while the polymer to porogen ratio is directly correlated to the amount of porosity of the final structure. After the polymer solution has been cast the solvent is allowed to fully evaporate, then the composite structure in the mold is immersed in a bath of a liquid suitable for dissolving the porogen: water in the case of sodium chloride, saccharose and gelatin or an aliphatic solvent like hexane for use with paraffin. Once the porogen has been fully dissolved, a porous structure is obtained. Other than the small thickness range that can be obtained, another drawback of SCPL lies in its use of organic solvents which must be fully removed to avoid any possible damage to the cells seeded on the scaffold.

Gas Foaming: To overcome the need to use organic solvents and solid porogens, a technique using gas as a porogen has been developed. First, disc-shaped structures made of the desired polymer are prepared by means of compression molding using a heated mold. The discs are then placed in a chamber where they are exposed to high pressure CO_2 for several days. The pressure inside the chamber is gradually restored to atmospheric levels. During this procedure the pores are formed by the carbon dioxide molecules that abandon the polymer, resulting in a sponge-like structure. The main problems resulting from such a technique are caused by the excessive heat used during compression molding (which prohibits the incorporation of any temperature labile material into the polymer matrix) and by the fact that the pores do not form an interconnected structure.

Emulsification/Freeze-Drying: This technique does not require the use of a solid porogen like SCPL. First, a synthetic polymer is dissolved into a suitable solvent (e.g. polylactic acid in dichloromethane) then water is added to the polymeric solution and the two liquids are mixed in order to obtain an emulsion. Before the two phases can separate, the emulsion is cast into a mold and quickly frozen by means of immersion into liquid nitrogen. The frozen emulsion is subsequently freeze-dried to remove the dispersed water and the solvent, thus leaving a solidified, porous polymeric structure. While emulsification and freeze-drying allow for a faster preparation when compared to SCPL (since it does not require a time consuming leaching step), it still requires the use of solvents. Moreover, pore size is relatively small and porosity is often irregular. Freeze-drying by itself is also a commonly employed technique for the fabrication of scaffolds. In particular, it is used to prepare collagen sponges: collagen is dissolved into acidic solutions of acetic acid or hydrochloric acid that are cast into a mold, frozen with liquid nitrogen and then lyophilized.

Thermally Induced Phase Separation (TIPS): Similar to the previous technique, this phase separation procedure requires the use of a solvent with a low melting point that is easy to sublime. For example dioxane could be used to dissolve polylactic acid, then phase separation is induced through the addition of a small quantity of water: a polymer-rich and a polymer-poor phase are formed. Following cooling below the solvent melting point and some days of vacuum-drying to sublime the solvent, a porous scaffold is obtained. Liquid-liquid phase separation presents the same drawbacks of emulsification/freeze-drying.

Electrospinning: A highly versatile technique that can be used to produce continuous fibers from submicrometre to nanometre diameters. In a typical electrospinning set-up, a solution is fed through a spinneret and a high voltage is applied to the tip. The buildup of electrostatic repulsion within the charged solution, causes it to eject a thin fibrous stream. A mounted collector plate or rod with an opposite or grounded charge draws in the continuous fibers, which arrive to form a highly porous network. The primary advantages of this technique are its simplicity and ease of variation. At a laboratory level, a typical electrospinning set-up only requires a high voltage power supply (up to 30 kV), a syringe, a flat tip needle and a conducting collector. By modifying variables such as the distance to collector, magnitude of applied voltage, or solution flow rate—researchers can dramatically change the overall scaffold architecture.

CAD/CAM Technologies: Because most of the above techniques are limited when it comes to the control of porosity and pore size, computer assisted design and manufacturing techniques have been introduced to tissue engineering. First, a three-dimensional structure is designed using CAD software. The porosity can be tailored using algorithms within the software. The scaffold is then realised by using ink-jet printing of polymer powders or through Fused Deposition Modelling of a polymer melt.

A 2011 study by El-Ayoubi et al. investigated "3D-plotting technique to produce (biocompatible and biodegradable) poly-L-Lactide macroporous scaffolds with two different pore sizes" via solid free-form fabrication (SSF) with computer-aided-design (CAD), to explore therapeutic articular cartilage replacement as an "alternative to conventional tissue repair". The study found the smaller the pore size paired with mechanical stress in a bioreactor (to induce in vivo-like conditions), the higher the cell viability in potential therapeutic functionality via decreasing recovery time and increasing transplant effectiveness.

Laser-Assisted BioPrinting (LaBP): In a 2012 study, Koch et al. focused on whether Laser-assisted BioPrinting (LaBP) can be used to build multicellular 3D patterns in natural matrix, and whether the generated constructs are functioning and forming tissue. LaBP arranges small volumes of living cell suspensions in set high-resolution patterns. The investigation was successful, the researchers foresee that "generated tissue constructs might be used for in vivo testing by implanting them into animal models" (14). As of this study, only human skin tissue has been synthesized, though researchers project that "by integrating further cell types (e.g. melanocytes, Schwann cells, hair follicle cells) into the printed cell construct, the behaviour of these cells in a 3D in vitro microenvironment similar to their natural one can be analyzed", useful for drug discovery and toxicology studies.

Assembly Methods

One of the continuing, persistent problems with tissue engineering is mass transport limitations. Engineered tissues generally lack an initial blood supply, thus making it difficult for any implanted cells to obtain sufficient oxygen and nutrients to survive, and/or function properly.

Self-Assembly

Self-assembly may play an important role here, both from the perspective of encapsulating cells and proteins, as well as creating scaffolds on the right physical scale for engineered tissue constructs and cellular ingrowth. The micromasonry is an prime technology to get cells grown in a lab to assemble into three-dimensional shapes. To break down tissue into single-cell building blocks, researchers have to dissolve the extracellular mortar that normally binds them together. But once that glue is removed, it's quite difficult to get cells to reassemble into the complex structures that make up our natural tissues. While cells aren't easily stackable, building blocks are. So the micromasonry starts with the encapsulation of living cells in polymer cubes. From there, the blocks self-assemble in any shape using templates.

Additive Manufacturing

It might be possible to print organs, or possibly entire organisms using additive manufacturing techniques. A recent innovative method of construction uses an ink-jet mechanism to print precise layers of cells in a matrix of thermoreversible gel. Endothelial cells, the cells that line blood vessels, have been printed in a set of stacked rings. When incubated, these fused into a tube.

The field of three-dimensional and highly accurate models of biological systems is pioneered by multiple projects and technologies including a rapid method for creating tissues and even whole organs involves a 3D printer that can print the scaffolding and cells layer by layer into a working tissue sample or organ. The device is presented in a TED talk by Dr. Anthony Atala, M.D. the Director of the Wake Forest Institute for Regenerative Medicine, and the W.H. Boyce Professor and Chair of the Department of Urology at Wake Forest University, in which a full bladder is printed on stage during the seminar and then presented to the crowd. This specific technology has been used to print a bladder for a young man that would have otherwise died without the availability of a transplant, and is anticipated to be able to print kidneys and perhaps livers in the future for transplantation and theoretically for toxicology and other biological studies as well.

Scaffolding

In 2013, using a 3-d scaffolding of Matrigel in various configurations, substantial pancreatic organoids was produced in vitro. Clusters of small numbers of cells proliferated into 40,000 cells within one week. The clusters transform into cells that make either digestive enzymes or hormones like insulin, self-organising into branched pancreatic organoids that resemble the pancreas.

The cells are sensitive to the environment, such as gel stiffness and contact with other cells. Individual cells do not thrive; a minimum of four proximate cells was required for subsequent organoid development. Modifications to the medium composition produced either hollow spheres mainly composed of pancreatic progenitors, or complex organoids that spontaneously undergo pancreatic morphogenesis and differentiation. Maintenance and expansion of pancreatic progenitors require active Notch and FGF signaling, recapitulating in vivo niche signaling interactions.

The organoids were seen as potentially offering mini-organs for drug testing and for spare insulin-producing cells.

Tissue Culture

In many cases, creation of functional tissues and biological structures *in vitro* requires extensive culturing to promote survival, growth and inducement of functionality. In general, the basic requirements of cells must be maintained in culture, which include oxygen, pH, humidity, temperature, nutrients and osmotic pressure maintenance.

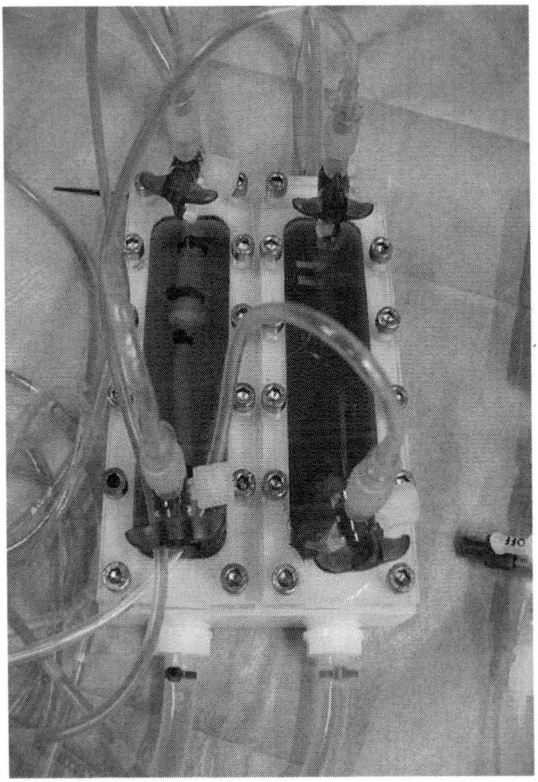

Figure: *Bioreactor for cultivation of vascular grafts*

Tissue engineered cultures also present additional problems in maintaining culture conditions. In standard cell culture, diffusion is often the sole means of nutrient and metabolite transport. However, as a culture becomes larger and more complex, such as the case with engineered organs and whole tissues, other mechanisms must be employed to maintain the culture, such as the creation of capillary networks within the tissue.

Another issue with tissue culture is introducing the proper factors or stimuli required to induce functionality. In many cases, simple maintenance culture is not sufficient. Growth factors, hormones, specific metabolites or nutrients, chemical and physical stimuli are sometimes required. For example, certain cells respond to changes in oxygen tension as part of their normal development, such as chondrocytes, which must adapt to low oxygen conditions or hypoxia during skeletal development. Others, such as endothelial cells, respond to shear stress from fluid flow, which is encountered in blood vessels. Mechanical stimuli, such as pressure pulses seem to be beneficial to all kind of cardiovascular tissue such as heart valves, blood vessels or pericardium.

Bioreactors

A bioreactor in tissue engineering, as opposed to industrial bioreactors, is a device that attempts to simulate a physiological environment in order to promote cell or tissue growth in vivo. A physiological environment can consist of many different parameters such as temperature and oxygen or carbon dioxide concentration, but can extend to all kinds of biological, chemical or mechanical stimuli. Therefore, there are systems that may include the application of forces or stresses to the tissue or even of electrical current in two- or three-dimensional setups.

In academic and industry research facilities, it is typical for bioreactors to be developed to replicate the specific physiological environment of the tissue being grown (e.g., flex and fluid shearing for heart tissue growth). Several general-use and application-specific bioreactors are also commercially available, and may provide static chemical stimulation or combination of chemical and mechanical stimulation.

The Bioreactors used for 3D cell cultures are small plastic cylindrical chambers with regulated internal humidity and moisture specifically engineered for the purpose of growing cells in three dimensions. The bioreactor uses bioactive synthetic materials such as polyethylene terephthalate membranes to surround the spheroid cells in an environment that maintains high levels of nutrients. They are easy to open and close, so that cell spheroids can be removed for testing, yet the chamber is able to maintain 100% humidity throughout. This humidity is important to achieve maximum cell growth and function. The bioreactor chamber is part of a larger device that rotates to ensure equal cell growth in each direction across three dimensions. MC2 Biotek has developed a bioreactor known as ProtoTissue that uses gas exchange to maintain high oxygen levels within the cell chamber; improving upon previous bioreactors, because the higher oxygen levels help the cell grow and undergo normal cell respiration.

Long Fiber Generation

In 2013, a group from the University of Tokyo developed cell laden fibers up to a metre in length and on the order of 100 μm in size. These fibers were created using a microfluidic device that forms a double coaxial laminar flow. Each 'layer' of the microfluidic device (cells seeded in ECM, a hydrogel sheath, and finally a calcium chloride solution). The seeded cells culture within the hydrogel sheath for several days, and then the sheath is removed with viable cell fibers. Various cell types were inserted into the ECM core, including myocytes, endothelial cells, nerve cell fibers, and epithelial cell fibers. This group then showed that these fibers can be woven together to fabricate tissues or organs in a mechanism similar to textile weaving. Fibrous morphologies are advantageous in that they provide an alternative to traditional scaffold design, and many organs (such as muscle) are composed of fibrous cells.

Bioartificial Organs

An article printed in the New York Times "A First: Organs Tailor-Made With Body's Own Cells," it recognised the first procedure of regenerative medicine of an implantation of a "bioartificial" organ. In this case it was a windpipe, that was suggested by a doctor at the Karolinska Institute in Sweden. This and the artificial bladder is just the start to what scientists and doctors want to achieve. Goals of making more complex organs such as the heart, liver, and kidneys are just some of few. Back in 2011 a patient was going to die unless he received a new windpipe. Dr. Macchiarini proposed it to his patient informing him that this procedure had only been tested on pigs. After slight hesitation and doubts he accepted. This procedure involved stem cells, but also a scaffold that could be used to make a whole artificial organ. The doctor extracted cells from the patients bone marrow and then with a synthetic scaffold they drip it "like basting a turkey" states this article in the New York Times. This continues until the human cells regenerate on their own and forms the organ needed. This kind of procedure helps eliminates some issues of a regular organ donated. This is "Because the donor's own cells are removed, this approach all but eliminates a major problem of transplants: the risk that foreign tissue will be rejected by the recipient." Dr. Macchiarini has more plans and goals however. He wishes to one day make organs without the synthetic scaffold which in the end is the main goal of most scientists and doctors working on tissue engineering.

Induced Stem Cells

Induced stem cells (iSC) are stem cells artificially derived from somatic, reproductive, pluripotent or other cell types by deliberate epigenetic reprogramming. They are classified as either totipotent (iTC), pluripotent (iPSC) or progenitor (multipotent—iMSC, also called an induced multipotent progenitor cell—iMPC) or unipotent — (iUSC) according to their developmental potential and degree of dedifferentiation. Progenitors are obtained by so-called direct reprogramming or directed differentiation and are also called induced somatic stem cells.

Three techniques are widely recognised:

- Transplantation of nuclei taken from somatic cells into a fertilized egg or oocyt from which the nucleus is removed prior
- Fusion of somatic cells with pluripotent stem cells and
- Modification of somatic cells, inducing its transformation into a stem cell, using: the genetic material encoding reprogramming protein factors, recombinant proteins; microRNA, a synthetic, self-replicating polycistronic RNA, and low-molecular weight biologically active substances.

Microfluidics

Microfluidics is a multidisciplinary field intersecting engineering, physics, chemistry, biochemistry, nanotechnology, and biotechnology, with practical applications to the design of systems in which small volumes of fluids will be handled. Microfluidics emerged in the beginning of the 1980s and is used in the development of inkjet printheads, DNA chips, lab-on-a-chip technology, micro-propulsion, and micro-thermal technologies. It deals with the behaviour, precise control and manipulation of fluids that are geometrically constrained to a small, typically sub-millimetre, scale. Typically, micro means one of the following features:

- small volumes (μL, nL, pL, fL)
- small size
- low energy consumption
- effects of the micro domain

Typically fluids are moved, mixed, separated or otherwise processed. Numerous applications employ passive fluid control techniques like capillary forces. In some applications external actuation means are additionally used for a directed transport of the media. Examples are rotary drives applying centrifugal forces for the fluid transport on the passive chips. Active microfluidics refers to the defined manipulation of the working fluid by active (micro) components as micropumps or micro valves. Micro pumps supply fluids in a continuous manner or are used for dosing. Micro valves determine the flow direction or the mode of movement of pumped liquids. Often processes which are normally carried out in a lab are miniaturized on a single chip in order to enhance efficiency and mobility as well as reducing sample and reagent volumes.

Microscale Behaviour of Fluids

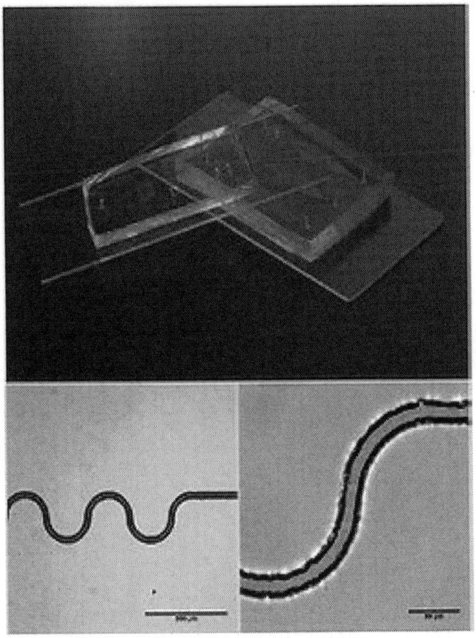

Figure: *Silicone rubber and glass microfluidic devices. Top: a photograph of the devices. Bottom: Phase contrast micrographs of a serpentine channel ~15 ìm wide.*

The behaviour of fluids at the microscale can differ from 'macrofluidic' behaviour in that factors such as surface tension, energy dissipation, and fluidic resistance start to dominate the system. Microfluidics studies how these behaviours change, and how they can be worked around, or exploited for new uses. At small scales (channel diameters of around 100 nanometres to several hundred micrometres) some interesting and sometimes unintuitive properties appear. In particular, the Reynolds number (which compares the effect of momentum of a fluid to the effect of viscosity) can become very low. A key consequence of this is that fluids, when side-by-side, do not necessarily mix in the traditional sense; molecular transport between them must often be through diffusion.

High specificity of chemical and physical properties (concentration, pH, temperature, shear force, etc.) can also be ensured resulting in more uniform reaction conditions and higher grade products in single and multi-step reactions.

Key Application Areas

Microfluidic structures include micropneumatic systems, i.e. microsystems for the handling of off-chip fluids (liquid pumps, gas valves, etc.), and microfluidic structures for the on-chip handling of nano- and picolitre volumes. To date, the most successful commercial application of microfluidics is the inkjet printhead. Significant research has been applied to the application of microfluidics for the production of industrially relevant quantities of material. Advances in microfluidics technology are revolutionizing molecular biology procedures for enzymatic analysis (e.g., glucose and lactate assays), DNA analysis (e.g., polymerase chain reaction and high-throughput sequencing), and proteomics. The basic idea of microfluidic biochips is to integrate assay operations such as detection, as well as sample pre-treatment and sample preparation on one chip.

An emerging application area for biochips is clinical pathology, especially the immediate point-of-care diagnosis of diseases. In addition, microfluidics-based devices, capable of continuous sampling and real-time testing of air/water samples for biochemical toxins and other dangerous pathogens, can serve as an always-on "bio-smoke alarm" for early warning.

Continuous-Flow Microfluidics

These technologies are based on the manipulation of continuous liquid flow through microfabricated channels. Actuation of liquid flow is implemented either by external pressure sources, external mechanical pumps, integrated mechanical micropumps, or by combinations of capillary forces and electrokinetic mechanisms. Continuous-flow microfluidic operation is the mainstream approach because it is easy to implement and less sensitive to protein fouling problems. Continuous-flow devices are adequate for many well-defined and simple biochemical applications, and for certain tasks such as chemical separation, but they are less suitable for tasks requiring a high degree of flexibility or ineffect fluid manipulations. These closed-channel systems are inherently difficult to integrate and scale because the parameters that govern flow field vary along the flow path making the fluid flow at any one location dependent on the properties of the entire system. Permanently etched microstructures also lead to limited reconfigurability and poor fault tolerance capability.

Process monitoring capabilities in continuous-flow systems can be achieved with highly sensitive microfluidic flow sensors based on MEMS technology which offer resolutions down to the nanoliter range.

Droplet-Based Microfluidics

Droplet-based microfluidics as a subcategory of microfluidics in contrast with continuous microfluidics has the distinction of manipulating discrete volumes of fluids in immiscible phases with low Reynolds number and laminar flow regimes. Interest in droplet-based microfluidics systems has been growing substantially in past decades. Microdroplets offer the feasibility of handling miniature volumes of fluids conveniently, provide better mixing and are suitable for high throughput experiments. Exploiting the benefits of droplet based microfluidics efficiently requires a deep understanding of droplet generation, droplet motion, droplet merging, and droplet breakup.

Digital Microfluidics

Alternatives to the above closed-channel continuous-flow systems include novel open structures, where discrete, independently controllable droplets are manipulated on a substrate using electrowetting. Following the analogy of digital microelectronics, this approach is referred to as digital microfluidics. Le Pesant et al. pioneered the use of electrocapillary forces to move droplets on a digital track. The "fluid transistor" pioneered by Cytonix also played a role. The technology was subsequently commercialized by Duke University. By using discrete unit-volume droplets, a microfluidic function can be reduced to a set of repeated basic operations, i.e., moving one unit of fluid over one unit of distance. This "digitization" method facilitates the use of a hierarchical and cell-based approach for microfluidic biochip design. Therefore, digital microfluidics offers a flexible and scalable system architecture as well as high fault-tolerance capability. Moreover, because each droplet can be controlled independently, these systems also have dynamic reconfigurability, whereby groups of unit cells in a microfluidic array can be reconfigured to change their functionality during the concurrent execution of a set of bioassays. Although droplets are manipulated in confined microfluidic channels, since the control on droplets is not independent, it should not be confused as "digital microfluidics". One common actuation method for digital microfluidics is electrowetting-on-dielectric (EWOD). Many lab-on-a-chip applications have been demonstrated within the digital microfluidics paradigm using electrowetting. However, recently other techniques for droplet manipulation have also been demonstrated using Surface Acoustic Waves, optoelectrowetting, mechanical actuation, etc.

DNA Chips (Microarrays)

Early biochips were based on the idea of a DNA microarray, e.g., the GeneChip DNAarray from Affymetrix, which is a piece of glass, plastic or silicon substrate on which pieces of DNA (probes) are affixed in a microscopic array. Similar to a DNA microarray, a protein array is a miniature array where a multitude of different capture agents, most frequently monoclonal antibodies, are deposited on a chip surface; they are used to determine the presence and/ or amount of proteins in biological samples, e.g., blood. A drawback of DNA and protein arrays is that they are neither reconfigurable nor scalable after manufacture. Digital microfluidics has been described as a means for carrying out Digital PCR.

Molecular Biology

In addition to microarrays biochips have been designed for two-dimensional electrophoresis, transcriptome analysis, and PCR amplification. Other applications include various electrophoresis and liquid chromatography applications for proteins and DNA, cell separation, in particular blood cell separation, protein analysis, cell manipulation and analysis including cell viability analysis and microorganism capturing.

Evolutionary Biology

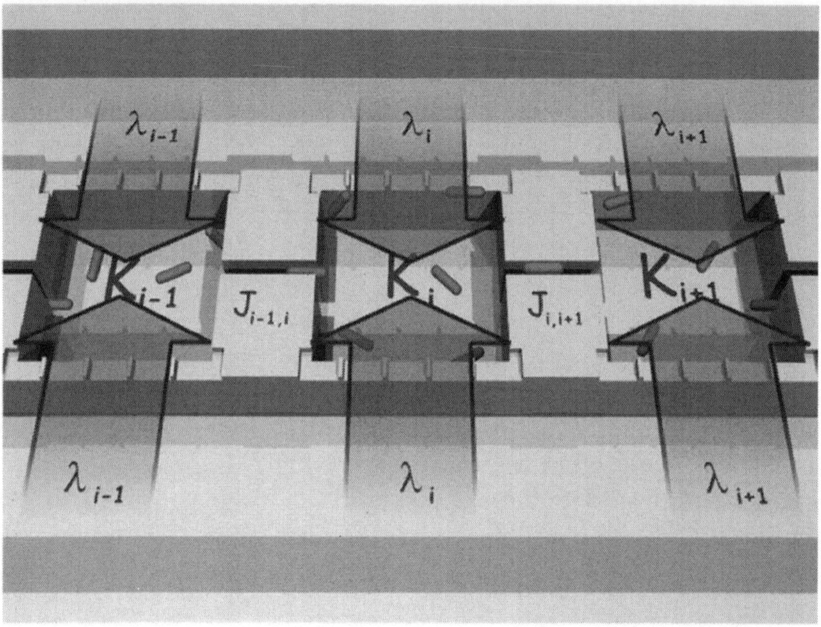

Figure: *Three Micro Habitat Patches MHPs connected by dispersal corridors (indicated here as* $J_{i,j}$*) into a 1D lattice. The ecosystem service (of habitat renewal) to each MHP represented here as* λ_i *(red arrows). Each MHP can also hold different carrying capacity* K_i *for its supporting local population of bacterial cells (depicted in green).*

By combining microfluidics with landscape ecology and nanofluidics, a nano/micro fabricated fluidic landscape can be constructed by building local patches of bacterial habitat and connecting them by dispersal corridors. The resulting landscapes can be used as physical implementations of an adaptive landscape, by generating a spatial mosaic of patches of opportunity distributed in space and time. The patchy nature of these fluidic landscapes allows for the study of adapting bacterial cells in a metapopulation system. The evolutionary ecology of these bacterial systems in these synthetic ecosystems allows for using biophysics to address questions in evolutionary biology.

Microbial Behaviour

The ability to create precise and carefully controlled chemoattractant gradients makes microfluidics the ideal tool to study motility, chemotaxis and the ability to evolve / develop resistance to antibiotics in small populations of microorganisms and in a short period of time. These microorganisms including bacteria and the broad range of organisms that form the marine microbial loop, responsible for regulating much of the oceans' biogeochemistry.

Cellular Biophysics

By rectifying the motion of individual swimming bacteria, microfluidic structures can be used to extract mechanical motion from a population of motile bacterial cells. This way, bacteria-powered rotors can be built.

Optics

The merger of microfluidics and optics is typical known as optofluidics. Examples of optofluidic devices :

Tuneable Microlens Array

Optofluidic Microscopes

Acoustic Droplet Ejection (ADE)

Acoustic droplet ejection uses a pulse of ultrasound to move low volumes of fluids (typically nanoliters or picoliters) without any physical contact. This technology focuses acoustic energy into a fluid sample in order to eject droplets as small as a millionth of a millionth of a liter (picoliter = 10^{-12} liter). ADE technology is a very gentle process, and it can be used to transfer proteins, high molecular weight DNA and live cells without damage or loss of viability. This feature makes the technology suitable for a wide variety of applications including proteomics and cell-based assays.

Fuel Cells

Microfluidic fuel cells can use laminar flow to separate the fuel and its oxidant to control the interaction of the two fluids without a physical barrier as would be required in conventional fuel cells.

Electroosmotic Pump

An electroosmotic pump (EOP), or EO pump, is used for generating flow or pressure by use of an electric field. One application of this is removing liquid flooding water from channels and gas diffusion layers and direct hydration of the proton exchange membrane in the membrane electrode assembly (MEA) of the proton exchange membrane fuel cells.

Electroosmotic pumps are fabricated from silica nanospheres or hydrophilic porous glass, the pumping mechanism is generated by an external electric field applied on an electric double layer (EDL), generates high pressures (more than 340 atm (34 MPa) at 12 kV applied potentials) and high flow rates (e.g., 40 ml/min at 100 V in a pumping structure less than 1 cm^3 in volume). EO pumps are compact, have no moving parts, and scale favourably with fuel cell design. The EO pump might drop the parasitic load of water management in fuel cells from 20% to 0.5% of the fuel cell power.

A Tool for Cell Biological Research

Microfluidic technology is creating powerful tools for cell biologists to control the complete cellular environment, leading to new questions and new discoveries. Many diverse advantages of this technology for microbiology are listed below:

- Single cell studies
- Microenvironmental control: ranging from mechanical environment to chemical environment
- Precise spatiotemporal concentration gradients
- Mechanical deformation
- Force measurements of adherent cells
- Confining cells
- Exerting a controlled force
- Fast and precise temperature control
- Electric field integration
- Cell culture
- Plant on a chip and plant tissue culture
- Antibiotic resistance: microfluidic devices can be used as heterogeneous environments for microorganisms. In an heterogeneous environment is easier for a microorganism to evolve. This can be useful for testing the acceleration of evolution of a microorganism / for testing the development of antibiotic resistance.

Future Directions

- On-chip characterization:
- Microfluidics in the classroom: On-chip acid-base titrations

Polymeric Nanoparticles for Targeted Release of Conventional and Macromolecular Drugs

Conventional small molecule drugs and large macromolecular drugs have significant and distinctly different delivery barriers. For example, small molecule drugs, such as the chemotherapeutic doxorubicin, is highly hydrophobic, thus administration requires toxic cosolvents to aid blood solubility. Macromolecular drugs, on the other hand, suffer from enzymatic degradation and inactivation, difficulty in targeting to the appropriate cells and transversing the cell membrane, and often become degraded intracellularly once endocytosed. We are investigating polymer-drug complexes or polymer-drug conjugates to overcome these barriers and modulate drug delivery.

Specifically, we synthesize polymers using reversible addition chain transfer polymerization (RAFT) techniques that enable the synthesis of many varied polymer architectures (linear, block copolymers, dendrimers, graft copolymers, gradient polymers), afford succinct control over molecular weight and polydispersity, and provide telechelic chain ends, amenable to further functionalization for drug conjugation or addition of cell/tissue targeting molecules.

Targeted drug delivery, sometimes called smart drug delivery, is a method of delivering medication to a patient in a manner that increases the concentration of the medication in some parts of the body relative to others. The goal of a targeted drug delivery system is to prolong, localize, target and have a protected drug interaction with the diseased tissue. The conventional drug delivery system is the absorption of the drug across a biological membrane, whereas the *targeted release system* releases the drug in a dosage form. The advantages to the targeted release system is the reduction in the frequency of the dosages taken by the patient, having a more uniform effect of the drug, reduction of drug side-effects, and reduced fluctuation in circulating drug levels. The disadvantage of the system is high cost, which makes productivity more difficult and the reduced ability to adjust the dosages.

Targeted drug delivery systems have been developed to optimize regenerative techniques. The system is based on a method that delivers a certain amount of a therapeutic agent for a prolonged period of time to a targeted diseased area within the body. This helps maintain the required plasma and tissue drug levels in the body, thereby preventing any damage to the healthy tissue via the drug. The drug delivery system is highly integrated and requires various disciplines, such as chemists, biologists, and engineers, to join forces to optimize this system.

In traditional drug delivery systems such as oral ingestion or intravascular injection, the medication is distributed throughout the body through the systemic blood circulation. For most therapeutic agents, only a small portion of the medication reaches the organ to be affected. Targeted drug delivery seeks to concentrate the medication in the tissues of interest while reducing the relative concentration of the medication in the remaining tissues. For example, by avoiding the host's defence mechanisms and inhibiting non-specific distribution in the liver and spleen, a system can reach the intended site of action in higher concentrations. Targeted delivery is believed to improve efficacy while reducing side-effects.

When implementing a targeted release system, the following design criteria for the system must be taken into account: the drug properties, side-effects of the drugs, the route taken for the delivery of the drug, the targeted site, and the disease.

Increasing developments to novel treatments requires a controlled microenvironment that is accomplished only through the implementation of therapeutic agents whose side-effects can be avoided with targeted drug delivery. Advances in the field of targeted drug delivery to cardiac tissue will be an integral component to regenerate cardiac tissue.

There are two kinds of targeted drug delivery: active targeted drug delivery, such as some antibody medications, and passive targeted drug delivery, such as the enhanced permeability and retention effect (EPR-effect).

Delivery Vehicles

There are different types of drug delivery vehicles, such as polymeric micelles, liposomes, lipoprotein-based drug carriers, nano-particle drug carriers, dendrimers, etc. An ideal drug delivery vehicle must be non-toxic, biocompatible, non-immunogenic, biodegradable, and must avoid recognition by the host's defence mechanisms.

Liposomes

The most common vehicle currently used for targeted drug delivery is the liposome. Liposomes are non-toxic, non-hemolytic, and non-immunogenic even upon repeated injections; they are biocompatible and biodegradable and can be designed to avoid clearance mechanisms (reticuloendothelial system (RES), renal clearance, chemical or enzymatic inactivation, etc.) Lipid-based, ligand-coated nanocarriers can store their payload in the hydrophobic shell or the hydrophilic interior depending on the nature of the drug/contrast agent being carried. The only problem to using liposomes in vivo is their immediate uptake and clearance by the RES system and their relatively low stability in vitro. To combat this, polyethylene glycol (PEG) can be added to the surface of the liposomes. Increasing the mole percent of PEG on the surface of the liposomes by 4-10% significantly increased circulation time in vivo from 200 to 1000 minutes.

Micelles and Dendrimers

Another type of drug delivery vehicle used is polymeric micelles. They are prepared from certain amphiphilic co-polymers consisting of both hydrophilic and hydrophobic monomer units. They can be used to carry drugs that have poor solubility. This method offers little in the terms of size control or function malleability. Techniques that utilise reactive polymers along with a hydrophobic additive to produce a larger micelle that create a range of sizes have been developed.

Dendrimers are also polymer-based delivery vehicles. They have a core that branches out in regular intervals to form a small, spherical, and very dense nanocarrier.

Biodegradable Particles

Biodegradable particles have the ability to target diseased tissue as well as deliver their payload as a controlled-release therapy. Biodegradable particles bearing ligands to P-selectin, endothelial selectin (E-selectin) and ICAM-1 have been found to adhere to inflamed endothelium. Therefore, the use of biodegradable particles can be also be used for cardiac tissue.

Artificial DNA Nanostructures

The success of DNA nanotechnology in constructing artificially designed nanostructures out of nucleic acids such as DNA, combined with the demonstration of systems for DNA computing, has led to speculation that artificial nucleic acid nanodevices can be used to target drug delivery based upon directly sensing its environment. These methods make use of DNA solely as a structural material and a chemical, and do not make use of its biological role as the carrier of genetic information. Nucleic acid logic circuits that could potentially be used as the core of a system that releases a drug only in response to a stimulus such as a specific mRNA have been demonstrated. In addition, a DNA "box" with a controllable lid has been synthesized using the DNA origami method. This structure could encapsulate a drug in its closed state, and open to release it only in response to a desired stimulus.

Applications

Targeted drug delivery can be used to treat many diseases, such as the cardiovascular diseases and diabetes. However, the most important application of targeted drug delivery is to treat cancerous tumors. The American Heart Association rates cardiovascular disease as the number one cause of death in the United States. Each year 1.5 million myocardial infarctions (MI), also known as heart attacks, occur in the United States, with 500,000 leading to deaths. The costs related to heart attacks exceed $60 billion per year. Therefore, there is a need to come up with an optimum recovery system. The key to solving this problem lies in the effective use of pharmaceutical drugs that can be targeted directly to the diseased tissue. This technique can help develop many more regenerative techniques to cure various diseases. The development of a number of regenerative strategies in recent years for *curing* heart disease represents a paradigm shift away from conventional approaches that aim to *manage* heart disease.

Stem cell therapy can be used to help regenerate myocardium tissue and return the contractile function of the heart by creating/supporting a microenvironment before the MI. Developments in targeted drug delivery to tumors have provided the groundwork for the burgeoning field of targeted drug delivery to cardiac tissue. Recent developments have shown that there are different endothelial surfaces in tumors, which has led to the concept of endothelial cell adhesion molecule-mediated targeted drug delivery to tumors.

Liposomes can be used as drug delivery for the treatment of tuberculosis. The traditional treatment for TB is akin to chemotherapy which is not overly effective, which may be due to the failure of chemotherapy to make a high enough concentration at the infection site. The liposome delivery system allows for better microphage penetration and better builds a concentration at the infection site. The delivery of the drugs works intravenously and by inhalation. Oral intake is not advised because the liposomes break down in the Gastrointestinal System.

Targeted Therapy

Targeted therapy or molecularly targeted therapy is a type of medication that blocks the growth of cancer cells by interfering with specific targeted molecules needed for carcinogenesis and tumor growth, rather than by simply interfering with all rapidly dividing cells (e.g. with traditional chemotherapy). Radiotherapy is not considered a 'targeted therapy' despite its often being aimed at the tumours.

Targeted cancer therapies are expected to be more effective than current treatments and less harmful to normal cells.

There are targeted therapies for breast cancer, multiple myeloma, lymphoma, prostate cancer, melanoma and other cancers.

The definitive experiments that showed that targeted therapy would reverse the malignant phenotype of tumor cells involved treating Her2/neu transformed cells with monoclonal antibodies in vitro and in vivo by Mark Greene's laboratory and reported from 1985.

Some have challenged use of the term, stating that drugs usually associated with the term are insufficiently selective. The phrase occasionally appears in scare quotes: "targeted therapy".

Types

The main categories of targeted therapy are *small molecules*, *small molecule drug conjugates* and *monoclonal antibodies.*

Small Molecules

- Imatinib mesylate (Gleevec, also known as STI–571) is approved for chronic myelogenous leukemia, gastrointestinal stromal tumor and some other types of cancer. Early clinical trials indicate that imatinib may be effective in treatment of dermatofibrosarcoma protuberans.
- Gefitinib (Iressa, also known as ZD1839), targets the epidermal growth factor receptor (EGFR) tyrosine kinase and is approved in the U.S. for non small cell lung cancer.
- Erlotinib (marketed as Tarceva). Erlotinib inhibits epidermal growth factor receptor, and works through a similar mechanism as gefitinib.

Erlotinib has been shown to increase survival in metastatic non small cell lung cancer when used as second line therapy. Because of this finding, erlotinib has replaced gefitinib in this setting.

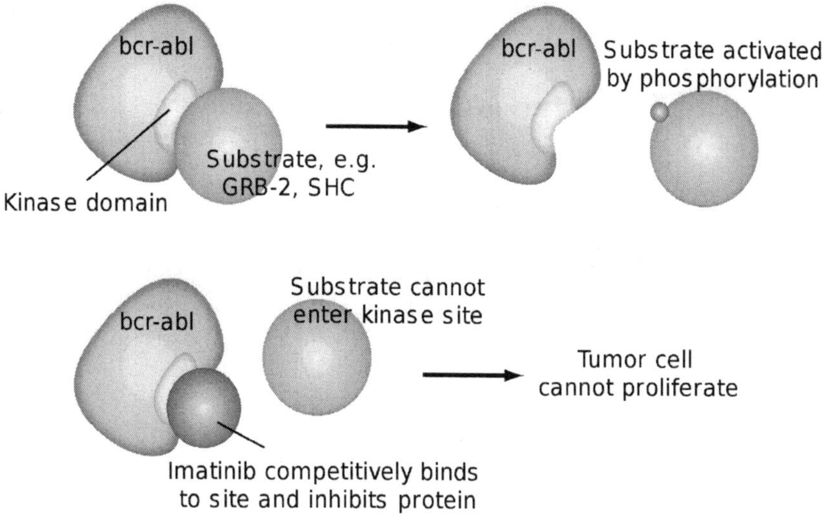

Figure: *Many are tyrosine-kinase inhibitors.*

- Bortezomib (Velcade) is an apoptosis-inducing proteasome inhibitor drug that causes cancer cells to undergo cell death by interfering with proteins. It is approved in the U.S. to treat multiple myeloma that has not responded to other treatments.
- The selective estrogen receptor modulator tamoxifen has been described as the foundation of targeted therapy.
- Janus kinase inhibitors, e.g. FDA approved tofacitinib
- ALK inhibitors, e.g. crizotinib
- Bcl-2 inhibitors (e.g. obatoclax in clinical trials, ABT-263, and Gossypol.
- PARP inhibitors (e.g. Iniparib, Olaparib in clinical trials)
- PI3K inhibitors (e.g. perifosine in a phase III trial)
- Apatinib is a selective VEGF Receptor 2 inhibitor which has shown encouraging anti-tumor activity in a broad range of malignancies in clinical trials. Apatinib is currently in clinical development for metastatic gastric carcinoma, metastatic breast cancer and advanced hepatocellular carcinoma.
- AN-152, (AEZS-108) doxorubicin linked to [D-Lys(6)]- LHRH, Phase II results for ovarian cancer.
- Braf inhibitors (vemurafenib, dabrafenib, LGX818) used to treat metastatic melanoma that harbors BRAF V600E mutation

- MEK inhibitors (trametinib, MEK162) are used in experiments, often in combination with BRAF inhibitors to treat melanoma
- CDK inhibitors, e.g. PD-0332991, LEE011 in clinical trials
- Hsp90 inhibitors, some in clinical trials
- salinomycin has demonstrated potency in killing cancer stem cells in both laboratory-created and naturally occurring breast tumors in mice.

Small Molecule Drug Conjugates

- Vintafolide is a small molecule drug conjugate consisting of a small molecule targeting the folate receptor. It is currently in clinical trials for platinum-resistant ovarian cancer (PROCEED trial) and a Phase 2b study(TARGET trial) in non-small-cell lung carcinoma (NSCLC).

Monoclonal Antibodies

Several are in development and a few have been licenced by the FDA. Examples of licenced monoclonal antibodies include:

- Rituximab (marketed as MabThera or Rituxan) targets CD20 found on B cells. It is used in non Hodgkin lymphoma
- Trastuzumab (Herceptin) targets the Her2/neu (also known as ErbB2) receptor expressed in some types of breast cancer
- Cetuximab (marketed as Erbitux) and Panitumumab target the epidermal growth factor receptor (EGFR). They are used in the treatment of colon cancer and non-small cell lung cancer.
- Bevacizumab (marketed as Avastin) targets circulating VEGF ligand. It is approved for use in the treatment of colon cancer, breast cancer, non-small cell lung cancer, and is investigational in the treatment of sarcoma. Its use for the treatment of brain tumors has been recommended.

Many Antibody-drug conjugates (ADCs) are being developed.

ADEPT (Antibody-Directed Enzyme Prodrug Therapy)

Directed Enzyme Prodrug Therapy (DEPT) uses enzymes artificially introduced into the body to convert Prodrugs, which have no or poor biological activity, to the active form in the desired location within the body.

Many chemotherapy drugs for cancer lack tumour specificity and the doses required to reach therapeutic levels in the tumour are often toxic to other tissues.

DEPT strategies are an experimental method of reducing the systemic toxicity of a drug, by achieving high levels of the active drug only at the desired site. This article describes the variations of DEPT technology.

Antibody-Directed Enzyme Prodrug Therapy (ADEPT)

ADEPT is a strategy to overcome the problems of lack of tumor selectivity. An antibody designed/developed against a tumor antigen is linked to an enzyme and injected to the blood, resulting in selective binding of the enzyme in the tumor. When the discrimination between tumor and normal tissue enzyme levels is sufficient, a prodrug is administrated into the blood circulation, which is converted to an active cytotoxic drug by the enzyme, only within the tumor. Selectivity is achieved by the tumor specificity of the antibody and by delaying prodrug administration until there is a large differential between tumor and normal tissue enzyme levels.

ADEPT has shown antitumor activity in animal tumor models of human choriocarcinoma and colonic and breast carcinoma.

ADEPT History

The first pilot-scale clinical trial of ADEPT was carried out at Charing Cross Hospital, London, using an anti-CEA F(ab2)2 antibody conjugated to the bacterial enzyme carboxypeptidase G2 (CPG2).

The antibody used in the first ADEPT clinical trial was of murine origin and the enzyme was bacterial. Host antibodies to both components of the AEC were present in the blood of all non-immunosuppressed patients by day 10 after AEC infusion. Several patients received ciclosporin since it had been shown in rabbits that this could delay the appearance of host antibodies to soluble proteins.

A subsequent, small-scale trial at the Royal Free Hospital, London, used the same agents as in the Charing Cross Hospital trial but the protocol was modified to provide additional pharmacokinetic data and most patients received only a single course of treatment.

Gene-Directed Enzyme Prodrug Therapy (GDEPT)

GDEPT is a suicide gene therapy in which the enzyme required for prodrug conversion is produced within the target cell, using a gene delivered to it by gene therapy. Systems that use viral vectors to deliver the gene are known as VDEPT.

Virus-Directed Enzyme Prodrug Therapy (VDEPT)

VDEPT is the term given to the use of a virus to deliver the gene for GDEPT. VDEPT can potentially be used to enhance the therapeutic potential of oncolytic viruses.

Polymer-Directed Enzyme Prodrug Therapy (PDEPT)

PDEPT uses polymer-drug conjugates, drugs contained within a polymer 'shell' such as pHPMA and designed to be released only by a specific enzyme.

Clostridia-Directed Enzyme Prodrug Therapy (CDEPT)

CDEPT is the use of Clostridia to convert prodrugs into active drug agents. CDEPT exploits the hypoxic environment of solid tumors to target drugs to tumors using anaerobic bacteria resident in the tumour to convert the prodrug to the active form. Intravenously injected clostridial spores exhibit a specificity for tumours, colonising the hypoxic areas of the tumours.

The CDEPT Strategy

Perhaps the most challenging issue in cancer treatment is how to reduce the side effects of the injected anti-cancer agents, which are of a high cytotoxicity potential. A widely used solution is to use enzymes which are able to convert a relatively non-toxic prodrug precursor into the active drug form(s). Clostridial-directed enzyme prodrug therapy (CDEPT) is one of the possible approaches.

Solid tumors, in contrast to normal tissues, grow rapidly. As a result, the cancerous tissues may suffer from inadequate blood and oxygen supply. Therefore, clostridia can grow in tumor and destroy it specifically. (Originally, Parker and co-workers showed that the injection of *Clostridium histolyticum* spores to the transplanted sarcomas of mice results in significant tumour lysis. Soon after, it was shown that a direct injection is not necessary, and that tumour colonization was readily obtained after intravenous administration of spores).

In CDEPT, a prodrug-converting enzyme expressed by a clostridial expression plasmid converts a prodrug into an active drug form within the tumor. While the prodrug is the inactive form and can be administrated to the blood, the products of the prodrug cleavage are highly cytotoxic and show their effect only in the vicinity of tumor cells.

Difficulties in the engineering of clostridial strains have restricted the application of other enzyme prodrug systems. So far, two enzymes have been applied in CDEPT: cytosine deaminase and nitroreductase. It has been recently suggested that β-lactamases, which are naturally found in Clostridia, can facilitate the application of the method significantly.

Progress and Future

In the U.S., the National Cancer Institute's *Molecular Targets Development Program* (MTDP) aims to identify and evaluate molecular targets that may be candidates for drug development.

8

Applications of Nanoparticles for Molecular Targeting and Selective Destruction of Cancer Cells

Breast cancer is a type of cancer originating from breast tissue, most commonly from the inner lining of milk ducts or the lobules that supply the ducts with milk. Cancers originating from ducts are known as ductal carcinomas, while those originating from lobules are known as lobular carcinomas. Breast cancer occurs in humans and other mammals. While the overwhelming majority of human cases are in women, breast cancer can also occur in men.

The balance of benefits versus harms of breast cancer screening is controversial. The characteristics of the cancer determine the treatment, which may include surgery, medications (hormonal therapy and chemotherapy), radiation and/or immunotherapy.

Surgery provides the single largest benefit, and to increase the likelihood of remission (no further sign of the cancer), several chemotherapy regimens are commonly given in addition. Radiation is used after breast-conserving surgery and substantially improves local relapse rates and in many circumstances also overall survival.

Worldwide, breast cancer accounts for 22.9% of all cancers (excluding non-melanoma skin cancers) in women. In 2008, breast cancer caused 458,503 deaths worldwide (13.7% of cancer deaths in women). Breast cancer is more than 100 times more common in women than in men, although men tend to have poorer outcomes due to delays in diagnosis.

Prognosis and survival rates for breast cancer vary greatly depending on the cancer type, stage, treatment, and geographical location of the patient. Survival rates in the Western world are high; for example, more than 8 out of 10 women (85%) in England diagnosed with breast cancer survive for at least 5 years. In developing countries, however, survival rates are much poorer.

Signs and Symptoms

The first noticeable symptom of breast cancer is typically a lump that feels different from the rest of the breast tissue. More than 80% of breast cancer cases are discovered when the woman feels a lump. The earliest breast cancers are detected by a mammogram. Lumps found in lymph nodes located in the armpits can also indicate breast cancer.

Indications of breast cancer other than a lump may include thickening different from the other breast tissue, one breast becoming larger or lower, a nipple changing position or shape or becoming inverted, skin puckering or dimpling, a rash on or around a nipple, discharge from nipple/s, constant pain in part of the breast or armpit, and swelling beneath the armpit or around the collarbone. Pain ("mastodynia") is an unreliable tool in determining the presence or absence of breast cancer, but may be indicative of other breast health issues.

Inflammatory breast cancer is a particular type of breast cancer which can pose a substantial diagnostic challenge. Symptoms may resemble a breast inflammation and may include itching, pain, swelling, nipple inversion, warmth and redness throughout the breast, as well as an orange-peel texture to the skin referred to as *peau d'orange*; as inflammatory breast cancer doesn't show as a lump there's sometimes a delay in diagnosis.

Another reported symptom complex of breast cancer is Paget's disease of the breast. This syndrome presents as skin changes resembling eczema, such as redness, discolouration, or mild flaking of the nipple skin.

As Paget's disease of the breast advances, symptoms may include tingling, itching, increased sensitivity, burning, and pain. There may also be discharge from the nipple. Approximately half of women diagnosed with Paget's disease of the breast also have a lump in the breast.

In rare cases, what initially appears as a fibroadenoma (hard, movable non-cancerous lump) could in fact be a phyllodes tumor. Phyllodes tumors are formed within the stroma (connective tissue) of the breast and contain glandular as well as stromal tissue. Phyllodes tumors are not staged in the usual sense; they are classified on the basis of their appearance under the microscope as benign, borderline, or malignant.

Occasionally, breast cancer presents as metastatic disease—that is, cancer that has spread beyond the original organ. The symptoms caused by metastatic breast cancer will depend on the location of metastasis. Common sites of metastasis include bone, liver, lung and brain.

Unexplained weight loss can occasionally herald an occult breast cancer, as can symptoms of fevers or chills. Bone or joint pains can sometimes be manifestations of metastatic breast cancer, as can jaundice or neurological symptoms. These symptoms are called *non-specific*, meaning they could be manifestations of many other illnesses.

Most symptoms of breast disorders, including most lumps, do not turn out to represent underlying breast cancer.

Fewer than 20% of lumps, for example, are cancerous, and benign breast diseases such as mastitis and fibroadenoma of the breast are more common causes of breast disorder symptoms.

Nevertheless, the appearance of a new symptom should be taken seriously by both patients and their doctors, because of the possibility of an underlying breast cancer at almost any age.

Risk Factors

The primary risk factors for breast cancer are female sex and older age. Other potential risk factors include: genetics, lack of childbearing or lack of breastfeeding, higher levels of certain hormones, certain dietary patterns, and obesity. Recent studies have indicated that exposure to light pollution is a risk factor for the development of breast cancer.

Lifestyle

Smoking tobacco appears to increase the risk of breast cancer, with the greater the amount smoked and the earlier in life that smoking began, the higher the risk. In those who are long-term smokers, the risk is increased 35% to 50%. A lack of physical activity has been linked to ~10% of cases.

The association between breast feeding and breast cancer has not been clearly determined; some studies have found support for an association while others have not. In the 1980s, the abortion–breast cancer hypothesis posited that induced abortion increased the risk of developing breast cancer.

This hypothesis was the subject of extensive scientific inquiry, which concluded that neither miscarriages nor abortions are associated with a heightened risk for breast cancer.

There may be an association between use of oral contraceptives and the development of premenopausal breast cancer, but whether oral contraceptives use may actually cause premenopausal breast cancer is a matter of debate. If there is indeed a link, the absolute effect is small.

In those with mutations in the breast cancer susceptibility genes *BRCA1* or *BRCA2*, or who have a family history of breast cancer, use of modern oral contraceptives does not appear to affect the risk of breast cancer.

There is a relationship between diet and breast cancer, including an increased risk with a high fat diet, alcohol intake, and obesity, related to higher cholesterol levels. Dietary iodine deficiency may also play a role.

Other risk factors include radiation, and shift-work. A number of chemicals have also been linked including: polychlorinated biphenyls, polycyclic aromatic hydrocarbons, organic solvents and a number of pesticides.

Although the radiation from mammography is a low dose, it is estimated that yearly screening from 40 to 80 years of age will cause approximately 225 cases of fatal breast cancer per million women screened.

Genetics

Some genetic susceptibility may play a minor role in most cases. Overall, however, genetics is believed to be the primary cause of 5–10% of all cases.

In those with zero, one or two affected relatives, the risk of breast cancer before the age of 80 is 7.8%, 13.3%, and 21.1% with a subsequent mortality from the disease of 2.3%, 4.2%, and 7.6% respectively.

In those with a first degree relative with the disease the risk of breast cancer between the age of 40 and 50 is double that of the general population.

In less than 5% of cases, genetics plays a more significant role by causing a hereditary breast–ovarian cancer syndrome. This includes those who carry the *BRCA1* and *BRCA2* gene mutation. These mutations account for up to 90% of the total genetic influence with a risk of breast cancer of 60–80% in those affected.

Other significant mutations include: *p53* (Li–Fraumeni syndrome), *PTEN* (Cowden syndrome), and *STK11* (Peutz–Jeghers syndrome), *CHEK2*, *ATM*, *BRIP1*, and *PALB2*.

In 2012, researchers said that there are four genetically distinct types of the breast cancer and that in each type, hallmark genetic changes lead to many cancers.

Medical Conditions

Breast changes like atypical ductal hyperplasia and lobular carcinoma *in situ*, found in benign breast conditions such as fibrocystic breast changes, are correlated with an increased breast cancer risk.

Diabetes mellitus might also increase the risk of breast cancer.

Pathophysiology

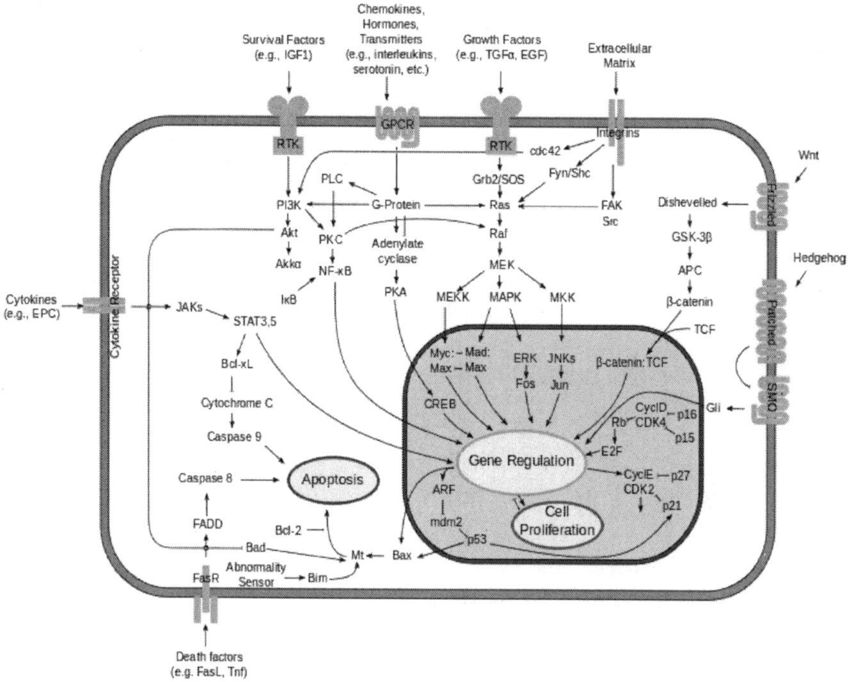

Figure: *Overview of signal transduction pathways involved in apoptosis. Mutations leading to loss of apoptosis can lead to tumorigenesis.*

Breast cancer, like other cancers, occurs because of an interaction between an environmental (external) factor and a genetically susceptible host. Normal cells divide as many times as needed and stop. They attach to other cells and stay in place in tissues. Cells become cancerous when they lose their ability to stop dividing, to attach to other cells, to stay where they belong, and to die at the proper time.

Normal cells will commit cell suicide (apoptosis) when they are no longer needed. Until then, they are protected from cell suicide by several protein clusters and pathways.

One of the protective pathways is the PI3K/AKT pathway; another is the RAS/MEK/ERK pathway. Sometimes the genes along these protective pathways are mutated in a way that turns them permanently "on", rendering the cell incapable of committing suicide when it is no longer needed.

This is one of the steps that causes cancer in combination with other mutations. Normally, the PTEN protein turns off the PI3K/AKT pathway when the cell is ready for cell suicide. In some breast cancers, the gene for the PTEN protein is mutated, so the PI3K/AKT pathway is stuck in the "on" position, and the cancer cell does not commit suicide.

Mutations that can lead to breast cancer have been experimentally linked to estrogen exposure.

Failure of immune surveillance, the removal of malignant cells throughout one's life by the immune system. Abnormal growth factor signaling in the interaction between stromal cells and epithelial cells can facilitate malignant cell growth. In breast adipose tissue, overexpression of leptin leads to increased cell proliferation and cancer.

In the United States, 10 to 20 percent of patients with breast cancer and patients with ovarian cancer have a first- or second-degree relative with one of these diseases. The familial tendency to develop these cancers is called hereditary breast–ovarian cancer syndrome.

The best known of these, the *BRCA* mutations, confer a lifetime risk of breast cancer of between 60 and 85 percent and a lifetime risk of ovarian cancer of between 15 and 40 percent. Some mutations associated with cancer, such as *p53*, *BRCA1* and *BRCA2*, occur in mechanisms to correct errors in DNA. These mutations are either inherited or acquired after birth. Presumably, they allow further mutations, which allow uncontrolled division, lack of attachment, and metastasis to distant organs.

However there is strong evidence of residual risk variation that goes well beyond hereditary *BRCA* gene mutations between carrier families. This is caused by unobserved risk factors. This implicates environmental and other causes as triggers for breast cancers. The inherited mutation in *BRCA1* or *BRCA2* genes can interfere with repair of DNA cross links and DNA double strand breaks (known functions of the encoded protein).

These carcinogens cause DNA damage such as DNA cross links and double strand breaks that often require repairs by pathways containing BRCA1 and BRCA2. However, mutations in *BRCA* genes account for only 2 to 3 percent of all breast cancers.

Levin *et al.* say that cancer may not be inevitable for all carriers of *BRCA1* and *BRCA2* mutations. About half of hereditary breast–ovarian cancer syndromes involve unknown genes.

GATA-3 directly controls the expression of estrogen receptor (ER) and other genes associated with epithelial differentiation, and the loss of GATA-3 leads to loss of differentiation and poor prognosis due to cancer cell invasion and metastasis.

Diagnosis

Most types of breast cancer are easy to diagnose by microscopic analysis of a sample—or biopsy—of the affected area of the breast. There are, however, rarer types of breast cancer that require specialised lab exams.

The two most commonly used screening methods, physical examination of the breasts by a healthcare provider and mammography, can offer an approximate likelihood that a lump is cancer, and may also detect some other lesions, such as a simple cyst.

When these examinations are inconclusive, a healthcare provider can remove a sample of the fluid in the lump for microscopic analysis (a procedure known as fine needle aspiration, or fine needle aspiration and cytology—FNAC) to help establish the diagnosis.

The needle aspiration may be performed in a healthcare provider's office or clinic using local anaesthetic if required. A finding of clear fluid makes the lump highly unlikely to be cancerous, but bloody fluid may be sent off for inspection under a microscope for cancerous cells. Together, physical examination of the breasts, mammography, and FNAC can be used to diagnose breast cancer with a good degree of accuracy.

Other options for biopsy include a core biopsy or vacuum-assisted breast biopsy, which are procedures in which a section of the breast lump is removed; or an excisional biopsy, in which the entire lump is removed. Very often the results of physical examination by a healthcare provider, mammography, and additional tests that may be performed in special circumstances (such as imaging by ultrasound or MRI) are sufficient to warrant excisional biopsy as the definitive diagnostic and primary treatment method.

Classification

Breast cancers are classified by several grading systems. Each of these influences the prognosis and can affect treatment response. Description of a breast cancer optimally includes all of these factors.

- Histopathology. Breast cancer is usually classified primarily by its histological appearance. Most breast cancers are derived from the epithelium lining the ducts or lobules, and these cancers are classified as ductal or lobular carcinoma. *Carcinoma in situ* is growth of low grade cancerous or precancerous cells within a particular tissue compartment such as the mammary duct without invasion of the surrounding tissue. In contrast, *invasive carcinoma* does not confine itself to the initial tissue compartment.
- Grade. Grading compares the appearance of the breast cancer cells to the appearance of normal breast tissue. Normal cells in an organ like the breast become differentiated, meaning that they take on specific shapes and forms that reflect their function as part of that organ. Cancerous cells lose that differentiation. In cancer, the cells that would normally line up in an orderly way to make up the milk ducts become

disorganised. Cell division becomes uncontrolled. Cell nuclei become less uniform. Pathologists describe cells as well differentiated (low grade), moderately differentiated (intermediate grade), and poorly differentiated (high grade) as the cells progressively lose the features seen in normal breast cells. Poorly differentiated cancers (the ones whose tissue is least like normal breast tissue) have a worse prognosis.

- Stage. Breast cancer staging using the TNM system is based on the size of the tumor (T), whether or not the tumor has spread to the lymph nodes (N) in the armpits, and whether the tumor has metastasized (M) (i.e. spread to a more distant part of the body). Larger size, nodal spread, and metastasis have a larger stage number and a worse prognosis.

The main stages are:

- o Stage 0 is a pre-cancerous or marker condition, either ductal carcinoma in situ (DCIS) or lobular carcinoma in situ (LCIS).
- o Stages 1–3 are within the breast or regional lymph nodes.
- o Stage 4 is 'metastatic' cancer that has a less favourable prognosis.

Where available, imaging studies may be employed as part of the staging process in select cases to look for signs of metastatic cancer. However, in cases of breast cancer with low risk for metastasis, the risks associated with PET scans, CT scans, or bone scans outweigh the possible benefits, as these procedures expose the patient to a substantial amount of potentially dangerous ionizing radiation.

- Receptor status. Breast cancer cells have receptors on their surface and in their cytoplasm and nucleus. Chemical messengers such as hormones bind to receptors, and this causes changes in the cell. Breast cancer cells may or may not have three important receptors: estrogen receptor (ER), progesterone receptor (PR), and HER2.

ER+ cancer cells (that is, cancer cells that have estrogen receptors) depend on estrogen for their growth, so they can be treated with drugs to block estrogen effects (e.g. tamoxifen), and generally have a better prognosis. Untreated, HER2+ breast cancers are generally more aggressive than HER2- breast cancers, but HER2+ cancer cells respond to drugs such as the monoclonal antibody trastuzumab (in combination with conventional chemotherapy), and this has improved the prognosis significantly.

Cells that do not have any of these three receptor types (estrogen receptors, progesterone receptors, or HER2) are called triple-negative, although they frequently do express receptors for other hormones, such as androgen receptor and prolactin receptor.

- DNA assays. DNA testing of various types including DNA microarrays have compared normal cells to breast cancer cells. The specific changes in a particular breast cancer can be used to classify the cancer in several ways, and may assist in choosing the most effective treatment for that DNA type.

Prevention

Women may reduce their risk of breast cancer by maintaining a healthy weight, drinking less alcohol, being physically active and breastfeeding their children. These modifications might prevent 38% of breast cancers in the US, 42% in the UK, 28% in Brazil and 20% in China. The benefits with moderate exercise such as brisk walking are seen at all age groups including postmenopausal women. Marine omega-3 polyunsaturated fatty acids appear to reduce the risk.

Removal of both breasts before any cancer has been diagnosed or any suspicious lump or other lesion has appeared (a procedure known as prophylactic bilateral mastectomy) may be considered in people with BRCA1 and BRCA2 mutations, which are associated with a substantially heightened risk for an eventual diagnosis of breast cancer. *BRCA* testing is recommended in those with a high family risk after genetic counseling. It is not recommended routinely. This is because there are many different forms of changes in *BRCA* genes, ranging from harmless polymorphisms to obviously dangerous frameshift mutations. The effect of most of identifiable changes in the genes is uncertain. Testing in an average-risk person is particularly likely to return one of these indeterminate, useless results.

The selective estrogen receptor modulators (such as tamoxifen) reduce the risk of breast cancer but increase the risk of thromboembolism and endometrial cancer. There is no overall change in the risk of death. They are thus not recommended for the prevention of breast cancer in women at average risk but may be offered for those at high risk. The benefit of breast cancer reduction continues for at least five years after stopping a course of treatment with these medications.

Screening

Breast cancer screening refers to testing otherwise-healthy women for breast cancer in an attempt to achieve an earlier diagnosis under the assumption that early detection will improve outcomes. A number of screening test have been employed including: clinical and self breast exams, mammography, genetic screening, ultrasound, and magnetic resonance imaging.

A clinical or self breast exam involves feeling the breast for lumps or other abnormalities. Clinical breast exams are performed by health care providers, while self breast exams are performed by the person themselves. Evidence does not support the effectiveness of either type of breast exam, as by the time a lump is large enough to be found it is likely to have been growing for several years and thus soon be large enough to be found without an exam.

Mammographic screening for breast cancer uses X-rays to examine the breast for any uncharacteristic masses or lumps. During a screening, the breast is compressed and a technician takes photos from multiple angles. A general mammogram takes photos of the entire breast, while a diagnostic mammogram focuses on a specific lump or area of concern.

A number of national bodies recommend breast cancer screening. For the average woman, the U.S. Preventive Services Task Force recommends mammography every two years in women between the ages of 50 and 74, the Council of Europe recommends mammography between 50 and 69 with most programs using a 2 year frequency, and in Canada screening is recommended between the ages of 50 and 74 at a frequency of 2 to 3 years. These task force reports point out that in addition to unnecessary surgery and anxiety, the risks of more frequent mammograms include a small but significant increase in breast cancer induced by radiation.

The Cochrane collaboration (2013) states that the best quality evidence neither demonstrates a reduction in cancer specific, nor a reduction in all cause mortality from screening mammography. When less rigorous trials are added to the analysis there is a reduction in mortality due to breast cancer of 0.05% (a decrease of 1 in 2000 deaths from breast cancer over 10 years or a relative decrease of 15% from breast cancer).

Screening over 10 years results in a 30% increase in rates of over-diagnosis and over-treatment (3 to 14 per 1000) and more than half will have at least one falsely positive test.

This has resulted in the view that it is not clear whether mammography screening does more good or harm. Cochrane states that, due to recent improvements in breast cancer treatment, and the risks of false positives from breast cancer screening leading to unnecessary treatment, "it therefore no longer seems reasonable to attend for breast cancer screening" at any age. Whether MRI as a screening method has greater harms or benefits when compared to standard mammography is not known.

Management

The management of breast cancer depends on various factors, including the stage of the cancer. Increasingly aggressive treatments are employed in accordance with the poorer the patient's prognosis and the higher the risk of recurrence of the cancer following treatment.

Breast cancer is usually treated with surgery, which may be followed by chemotherapy or radiation therapy, or both. A multidisciplinary approach is preferable. Hormone receptor-positive cancers are often treated with hormone-blocking therapy over courses of several years. Monoclonal antibodies, or other immune-modulating treatments, may be administered in certain cases of metastatic and other advanced stages of breast cancer.

Drugs used after and in addition to surgery are called adjuvant therapy. Chemotherapy or other types of therapy prior to surgery are called neoadjuvant therapy. Aspirin may reduce mortality from breast cancer.

There are currently three main groups of medications used for adjuvant breast cancer treatment: hormone-blocking agents, chemotherapy, and monoclonal antibodies.

Hormone Blocking Therapy

Some breast cancers require estrogen to continue growing. They can be identified by the presence of estrogen receptors (ER+) and progesterone receptors (PR+) on their surface (sometimes referred to together as hormone receptors). These ER+ cancers can be treated with drugs that either block the receptors, e.g. tamoxifen (Nolvadex), or alternatively block the production of estrogen with an aromatase inhibitor, e.g. anastrozole (Arimidex) or letrozole (Femara). Aromatase inhibitors, however, are only suitable for post-menopausal patients. This is because the active aromatase in postmenopausal women is different from the prevalent form in premenopausal women, and therefore these agents are ineffective in inhibiting the predominant aromatase of premenopausal women.

Chemotherapy

Chemotherapy is predominantly used for cases of breast cancer in stages 2–4, and is particularly beneficial in estrogen receptor-negative (ER-) disease. The chemotherapy medications are administered in combinations, usually for periods of 3–6 months.

One of the most common regimens, known as "AC", combines cyclophosphamide with doxorubicin (Adriamycin). Sometimes a taxane drug, such as docetaxel (Taxotere), is added, and the regime is then known as "CAT". Another common treatment, which produces equivalent results, is cyclophosphamide, methotrexate, and fluorouracil (or "CMF"). Most chemotherapy medications work by destroying fast-growing and/or fast-replicating cancer cells, either by causing DNA damage upon replication or by other mechanisms.

However, the medications also damage fast-growing normal cells, which may cause serious side effects. Damage to the heart muscle is the most dangerous complication of doxorubicin, for example.

Monoclonal Antibodies

Trastuzumab (Herceptin), a monoclonal antibody to HER2 (a cell receptor that is especially active in some breast cancer cells), has improved the 5-year disease free survival of stage 1–3 HER2-positive breast cancers to about 87% (overall survival 95%). When stimulated by certain growth factors, HER2 causes cellular growth and division; in the absence of stimulation by the growth factor, the cell will normally stop growing.

Between 25% and 30% of breast cancers overexpress the HER2 gene or its protein product, and overexpression of HER2 in breast cancer is associated with increased disease recurrence and worse prognosis. When trastuzumab binds to the HER2 in breast cancer cells that overexpress the receptor, trastuzumab prevents growth factors from being able to bind to and stimulate the receptors, effectively blocking the growth of the cancer cells.

Trastuzumab, however, is very expensive, and its use may cause serious side effects (approximately 2% of patients who receive it suffer significant heart damage). Further, trastuzumab is only effective in patients with HER2 amplification/overexpression.

Other monoclonal antibodies are also being evaluated in clinical trials.

Radiation

Radiotherapy is given after surgery to the region of the tumor bed and regional lymph nodes, to destroy microscopic tumor cells that may have escaped surgery. It may also have a beneficial effect on tumor microenvironment. Radiation therapy can be delivered as external beam radiotherapy or as brachytherapy (internal radiotherapy).

Conventionally radiotherapy is given *after* the operation for breast cancer. Radiation can also be given at the time of operation on the breast cancer-intraoperatively. The largest randomised trial to test this approach was the TAR-GIT-A Trial which found that targeted intraoperative radiotherapy was equally effective at 4-years as the usual several weeks' of whole breast external beam radiotherapy. Radiation can reduce the risk of recurrence by 50–66% (1/2 – 2/3 reduction of risk) when delivered in the correct dose and is considered essential when breast cancer is treated by removing only the lump (Lumpectomy or Wide local excision).

Prognosis

Prognosis is the long term outcomes of the condition. This includes the probability of progression-free survival (PFS) or disease-free survival (DFS). These predictions are based on experience with breast cancer patients with similar classification. A prognosis is an estimate, as patients with the same classification will survive a different amount of time, and classifications are not always precise. Survival is usually calculated as an average number of months (or years) that 50% of patients survive, or the percentage of patients that are alive after 1, 5, 15, and 20 years.

Prognosis is important for treatment decisions because patients with a good prognosis are usually offered less invasive treatments, such as lumpectomy and radiation or hormone therapy, while patients with poor prognosis are usually offered more aggressive treatment, such as more extensive mastectomy and one or more chemotherapy drugs.

Prognostic Factors

Prognostic factors are reflected in the classification scheme for breast cancer including stage, (i.e., tumor size, location, whether disease has spread to lymph nodes and other parts of the body), grade, recurrence of the disease, and the age and health of the patient. The Nottingham Prognostic Index is a commonly used prognostic tool.

The stage of the breast cancer is the most important component of traditional classification methods of breast cancer, because it has a greater effect on the prognosis than the other considerations. Staging takes into consideration size, local involvement, lymph node status and whether metastatic disease is present. The higher the stage at diagnosis, the poorer the prognosis.

The stage is raised by the invasiveness of disease to lymph nodes, chest wall, skin or beyond, and the aggressiveness of the cancer cells. The stage is lowered by the presence of cancer-free zones and close-to-normal cell behaviour (grading). Size is not a factor in staging unless the cancer is invasive. For example, Ductal Carcinoma In Situ (DCIS) involving the entire breast will still be stage zero and consequently an excellent prognosis with a 10-year disease free survival of about 98%.

- Stage 1 cancers (and DCIS, LCIS) have an excellent prognosis and are generally treated with lumpectomy and sometimes radiation. HER2+ cancers should be treated with the trastuzumab (Herceptin) regime. Chemotherapy is uncommon for other types of stage 1 cancers.
- Stage 2 and 3 cancers with a progressively poorer prognosis and greater risk of recurrence are generally treated with surgery (lumpectomy or mastectomy with or without lymph node removal), chemotherapy (plus trastuzumab for HER2+ cancers) and sometimes radiation (particularly following large cancers, multiple positive nodes or lumpectomy).
- Stage 4, metastatic cancer, (i.e. spread to distant sites) has poor prognosis and is managed by various combination of all treatments from surgery, radiation, chemotherapy and targeted therapies. 10-year survival rate is 5% without treatment and 10% with optimal treatment.

The breast cancer grade is assessed by comparison of the breast cancer cells to normal breast cells.

The closer to normal the cancer cells are, the slower their growth and the better the prognosis. If cells are not well differentiated, they will appear immature, will divide more rapidly, and will tend to spread. Well differentiated is given a grade of 1, moderate is grade 2, while poor or undifferentiated is given a higher grade of 3 or 4 (depending upon the scale used). The most widely used grading system is the Nottingham scheme; details are provided in the discussion of breast cancer grade.

The presence of estrogen and progesterone receptors in the cancer cell is important in guiding treatment. Those who do not test positive for these specific receptors will not be able to respond to hormone therapy, and this can affect their chance of survival depending upon what treatment options remain, the exact type of the cancer, and how advanced the disease is.

In addition to hormone receptors, there are other cell surface proteins that may affect prognosis and treatment. HER2 status directs the course of treatment. Patients whose cancer cells are positive for HER2 have more aggressive disease and may be treated with the 'targeted therapy', trastuzumab (Herceptin), a monoclonal antibody that targets this protein and improves the prognosis significantly.

Younger women tend to have a poorer prognosis than post-menopausal women due to several factors. Their breasts may change with their menstrual cycles, they may be nursing infants, and they may be unaware of changes in their breasts.

Therefore, younger women are usually at a more advanced stage when diagnosed. There may also be biologic factors contributing to a higher risk of disease recurrence for younger women with breast cancer.

High mammographic breast density, which is a marker of increased risk of developing breast cancer, may not mean an increased risk of death among breast cancer patients, according to a 2012 report of a study involving 9232 women by the National Cancer Institute (NCI).

Targeted Viral Therapy Destroys Breast Cancer Stem Cells in Preclinical Experiments

Virotherapy is a treatment using biotechnology to convert viruses into therapeutic agents by reprogramming viruses to treat diseases. There are three main branches of virotherapy: anti-cancer oncolytic viruses, viral vectors for gene therapy and viral immunotherapy. In a slightly different context, virotherapy can also refer more broadly to the use of viruses to treat certain medical conditions by killing pathogens.

Oncolytic Virotherapy

Oncolytic virotherapy is not a new idea – as early as the mid 1950s doctors were noticing that cancer patients who suffered a non-related viral infection, or who had been vaccinated recently, showed signs of improvement; this has been largely attributed to the production of interferon and tumour necrosis factors in response to viral infection, but oncolytic viruses are being designed that selectively target and lyse only cancerous cells.

In the 1940s and 1950s, studies were conducted in animal models to evaluate the use of viruses in the treatment of tumours.

In the 1940s-50s some of the earliest human clinical trials with oncolytic viruses were started. However, for several years research in this field was delayed due to the inadequate technology available. Research has now started to proceed more quickly in finding ways to use viruses therapeutically.

As well as the direct anti-cancer effect, oncolytic viruses are also capable of inducing an anti-tumour immune response.

Viral Gene Therapy

Viral gene therapy most frequently uses non-replicating viruses to deliver therapeutic genes to cells with genetic malfunctions.

Early efforts while technically successful, faced considerable delays due to safety issues as the uncontrolled delivery of a gene into a host genome has the potential to disrupt tumour suppressing genes and induce cancer, and did so in two cases. Immune responses to viral therapies also pose a barrier to successful treatment, for this reason eye therapy for genetic blindness is attractive as the eye is an immune privileged site, preventing an immune response.

An alternative form of viral gene therapy is to deliver a gene which may be helpful in preventing disease that would not normally be expressed in the natural disease condition. For example the growth of new blood vessels in cancer, known as angiogenesis, enables tumours to grow larger. However, a virus introducing anti-angiogenic factors to the tumour may be able to slow or halt growth.

Viral Immunotherapy

Viral immunotherapy uses viruses to introduce specific antigens to the patient's immune system. Unlike traditional vaccines, in which attenuated or killed virus/bacteria is used to generate an immune response, viral immunotherapy uses genetically engineered viruses to present a specific antigen to the immune system.

That antigen could be from any species of virus/bactera or even human disease antigens, for example cancer antigens.

Specific Projects and Products

Oncolytic Viruses

The immune modulating virotherapy known as RIGVIR developed at the Institute of Microbiology in Latvia is the first virotherapy to pass all stages of clinical development and was approved in 2009. Since 2004 Rigvir is approved in Latvia (national registration), but since 2008 Rigvir is available in pharmacies of Latvia. Right now virotherapy with Rigvir is successfully used in Latvia and 25 other countries around the world.

In 2004, researchers from University of Texas genetically programmed a type of common cold virus Adenovirus Delta-24-RGD to attack glioblastoma multiforme. Later other researchers have tried tests on mice where 9 out of 10 mice have shown degeneration of tumours and prolonged survival. A drug grade virus was approved for clinical trials on humans in 2009.

In 2006 researchers from the Hebrew University succeeded in isolating a variant of the Newcastle disease Virus (NDV-HUJ), which usually affects birds, in order to specifically target cancer cells. The researchers tested the new virotherapy on patients with glioblastoma multiforme and achieved promising results for the first time.

Vaccinia virus, a virus credited for the eradication of smallpox, is being developed as an oncolytic virus, e.g. GL-ONC1 and JX-594. Promising research results warrant its clinical development in human patients.

The experimental virotherapy that has progressed the farthest in clinical trials (as of 2013) is Talimogene laherparepvec. It is based on an engineered version of herpes simplex virus which has also been engineered to express GM-CSF. This virus is being developed by Amgen who reported that a pivotal phase 3 study in melanoma had met its primary endpoint (durable response rate) with a very high degree of significance in March 2013, the first positive phase 3 study with an oncolytic virus in the western world.

Viral Gene Therapy

ProSavin is one of a number of therapies in the Lentivector platform under development by Oxford BioMedica. It delivers to the brain the genes for three enzymes important in the production of dopamine, a deficiency of which causes Parkinson's disease.

TNFerade (a non replicating TNF gene therapy virus) failed a phase III trial for pancreatic cancer.

Viral Immunotherapy

Trovax is an immunotherapy that uses a pox-virus bearing the tumour antigen 5T4, to induce an immune response against a variety of cancer types.

The therapy was developed by Oxford BioMedica and failed to improve overall survival in a phase 3 trial in renal cell carcinoma. New phase II trials have since begun at Cardiff University (UK) with colorectal cancer and at the Velindre Cancer Centre (Cardiff, UK) with malignant pleural mesothelioma.

Cancer Stem Cell

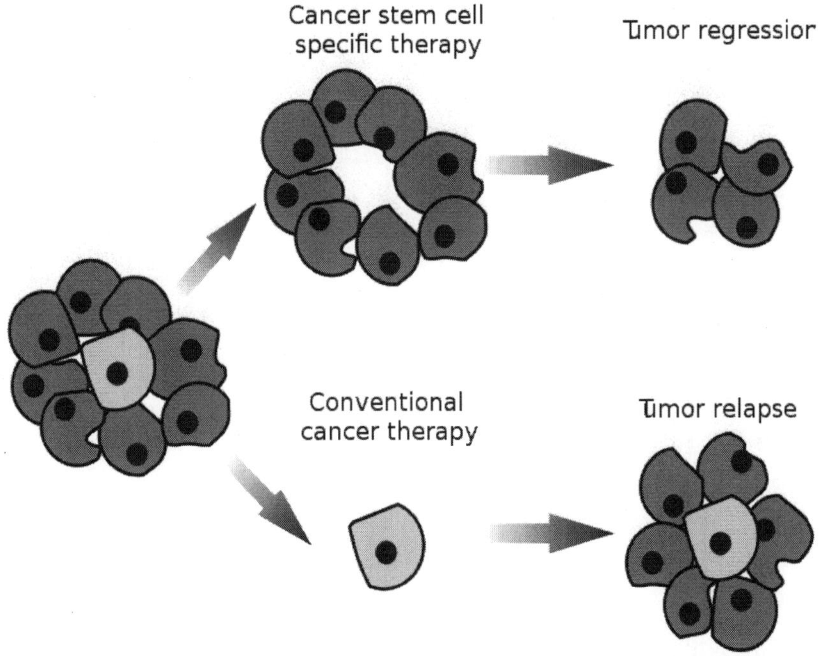

Figure: *Stem cell specific and conventional cancer therapies*

Cancer stem cells (CSCs) are cancer cells (found within tumors or hematological cancers) that possess characteristics associated with normal stem cells, specifically the ability to give rise to all cell types found in a particular cancer sample. CSCs are therefore tumorigenic (tumor-forming), perhaps in contrast to other non-tumorigenic cancer cells. CSCs may generate tumors through the stem cell processes of self-renewal and differentiation into multiple cell types. Such cells are proposed to persist in tumors as a distinct population and cause relapse and metastasis by giving rise to new tumors. Therefore, development of specific therapies targeted at CSCs holds hope for improvement of survival and quality of life of cancer patients, especially for sufferers of metastatic disease.

Existing cancer treatments have mostly been developed based on animal models, where therapies able to promote tumor shrinkage were deemed effective. However, animals could not provide a complete model of human disease. In particular, in mice, whose life spans do not exceed two years, tumor relapse is exceptionally difficult to study.

The efficacy of cancer treatments is, in the initial stages of testing, often measured by the ablation fraction of tumor mass (fractional kill). As CSCs would form a very small proportion of the tumor, this may not necessarily select for drugs that act specifically on the stem cells. The theory suggests that conventional chemotherapies kill differentiated or differentiating cells, which form the bulk of the tumor but are unable to generate new cells. A population of CSCs, which gave rise to it, could remain untouched and cause a relapse of the disease.

Models for Tumor Propagation

In different tumor subtypes, cells within the tumor population exhibit functional heterogeneity, and tumors are formed from cells with various proliferative and differentiate capacities. This functional tumour heterogeneity among cancer cells has led to the creation of at least two models, which have been put forward to account for heterogeneity and differences in tumor-regenerative capacity: the cancer stem cell (CSC) and clonal evolution models.

The cancer stem cell model refers to a subset of tumor cells that have the ability to self-renew and are able to generate the diverse tumor cells. These cells have been termed cancer stem cells to reflect their stem-like properties. One implication of the CSC model and the existence of CSCs is that the tumor population is hierarchically arranged with CSCs lying at the apex of the hierarchy.

Evidence of CSCs

The existence of CSCs is a subject of debate within medical research, because many studies have not been successful in discovering the similarities and differences between normal tissue stem cells and cancer (stem) cells. Cancer cells must be capable of continuous proliferation and self-renewal in order to retain the many mutations required for carcinogenesis, and to sustain the growth of a tumor since differentiated cells (constrained by the Hayflick Limit) cannot divide indefinitely. However, it is debated whether such cells represent a minority. If most cells of the tumor are endowed with stem cell properties, there is no incentive to focus on a specific subpopulation. There is also debate on the cell of origin of CSCs - whether they originate from normal stem cells that have lost the ability to regulate proliferation, or from more differentiated population of progenitor cells that have acquired abilities to self-renew (which is related to the issue of stem cell plasticity).

The first conclusive evidence for CSCs was published in 1997 in Nature Medicine. Bonnet and Dick isolated a subpopulation of leukaemic cells that expressed a specific surface marker CD34, but lacked the CD38 marker. The authors established that the $CD34^{+}/CD38^{-}$ subpopulation is capable of initiating tumors in NOD/SCID mice that are histologically similar to the donor. The first evidence of a solid tumor cancer stem-like cell followed in 2002 with the discovery of a clonogenic, sphere-forming cell isolated and characterized from human brain gliomas [Human cortical glial tumors contain neural stem-like cells expressing astroglial and neuronal markers in vitro.

In cancer research experiments, tumor cells are sometimes injected into an experimental animal to establish a tumor. Disease progression is then followed in time and novel drugs can be tested for their ability to inhibit it. However, efficient tumor formation requires thousands or tens of thousands of cells to be introduced.

Classically, this has been explained by poor methodology (i.e. the tumor cells lose their viability during transfer) or the critical importance of the microenvironment, the particular biochemical surroundings of the injected cells. Supporters of the CSC paradigm argue that only a small fraction of the injected cells, the CSCs, have the potential to generate a tumor. In human acute myeloid leukemia the frequency of these cells is less than 1 in 10,000.

Further evidence comes from histology, the study of the tissue structure of tumors. Many tumors are very heterogeneous and contain multiple cell types native to the host organ. Heterogeneity is commonly retained by tumor metastases. This implies that the cell that produced them had the capacity to generate multiple cell types. In other words, it possessed multidifferentiative potential, a classical hallmark of stem cells.

The existence of leukaemic stem cells prompted further research into other types of cancer. CSCs have recently been identified in several solid tumors, including cancers of the:

- Brain
- Breast
- Colon
- Ovary
- Pancreas
- Prostate
- Melanoma
- Multiple Myeloma

Mechanistic and Mathematical Models

Once the pathways to cancer are hypothesized, it is possible to develop predictive mathematical biology models, e.g., based on the cell compartment method. For instance, the growths of the abnormal cells from their normal counterparts can be denoted with specific mutation probabilities. Such a model has been employed to predict that repeated insult to mature cells increases the formation of abnormal progeny, and hence the risk of cancer. Considerable work needs to be done, however, before the clinical efficacy of such models is established.

Origin of CSC

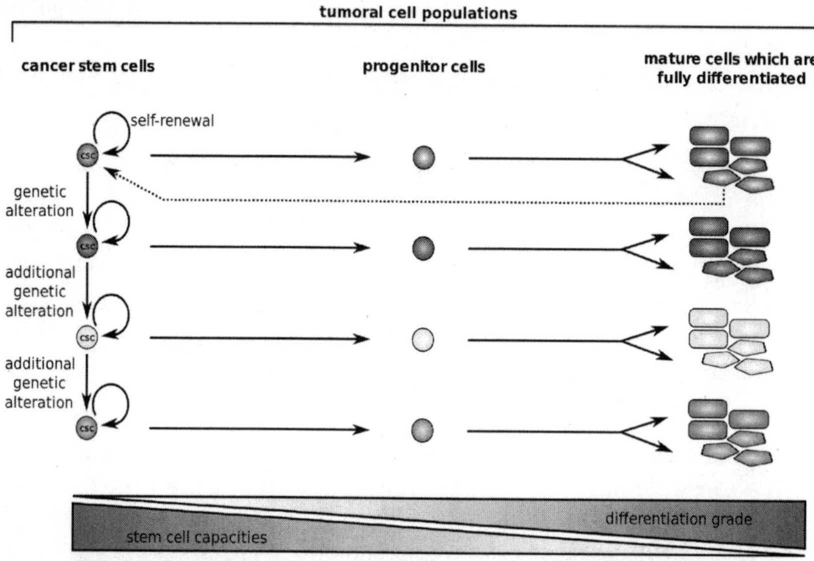

Figure: *Hierarchical organisation of a tumour according to the CSC model*

The origin of cancer stem cells is still an area of ongoing research. Several camps have formed within the scientific community regarding the issue, and it is possible that several answers are correct, depending on the tumor type and the phenotype the tumor presents.

One important distinction that will often be raised is that the cell of origin for a tumor can not be demonstrated using the cancer stem cell as a model. This is because cancer stem cells are isolated from end-stage tumors. Therefore, describing a cancer stem cell as a cell of origin is often an inaccurate claim, even though a cancer stem cell is capable of initiating new tumor formation.

With that caveat mentioned, various theories define the origin of cancer stem cells. In brief, CSC can be generated as: mutants in developing stem or progenitor cells, mutants in adult stem cells or adult progenitor cells, or mutant differentiated cells that acquire stem like attributes. These theories often do focus on a tumor's cell of origin and as such must be approached with skepticism.

Some researchers favour the theory that the cancer stem cell is generated by a mutation in stem cell niche populations during development. The logical progression claims that these developing stem populations are mutated and then expand such that the mutation is shared by many of the descendants of the mutated stem cell. These daughter stem cells are then much closer to becoming tumors, and since there are many of them there is more chance of a mutation that can cause cancer.

Another theory associates adult stem cells with the formation of tumors. This is most often associated with tissues with a high rate of cell turnover (such as the skin or gut). In these tissues, it has long been expected that stem cells are responsible for tumor formation.

This is a consequence of the frequent cell divisions of these stem cells (compared to most adult stem cells) in conjunction with the extremely long lifespan of adult stem cells.

This combination creates the ideal set of circumstances for mutations to accumulate; accumulation of mutations is the primary factor that drives cancer initiation. In spite of the logical backing of the theory, only recently has any evidence appeared showing association represents an actual phenomenon.

It is important to bear in mind that due to the heterogeneous nature of evidence it is possible that any individual cancer could come from an alternative origin.

A third possibility often raised is the potential de-differentiation of mutated cells such that these cells acquire stem cell like characteristics. This is often used as a potential alternative to any specific cell of origin, as it suggests that any cell might become a cancer stem cell.

Another related concept is the concept of tumor hierarchy. This concept claims that a tumor is a heterogeneous population of mutant cells, all of which share some mutations but vary in specific phenotype. In this model, the tumor is made up of several types of stem cells, one optimal to the specific environment and several less successful lines.

These secondary lines can become more successful in some environments, allowing the tumor to adapt to its environment, including adaptation to tumor treatment. If this situation is accurate, it has severe repercussions on cancer stem cell specific treatment regime. Within a tumor hierarchy model, it would be extremely difficult to pinpoint the cancer stem cell's origin.

Cancer Stem Cell Isolation

CSC, now reported in most human tumors, are commonly identified and enriched using strategies for identifying normal stem cells that are similar across various studies.

The procedures include fluorescence-activated cell sorting (FACS), with antibodies directed at cell-surface markers, and functional approaches including SP analysis (side population assay) or Aldefluor assay. The CSC-enriched population purified by these approaches is then implanted, at various cell doses, in immune-deficient mice to assess its tumor development capacity. This *in vivo* assay is called limiting dilution assay.

The tumor cell subsets that can initiate tumor development at low cell numbers are further tested for self-renewal capacity in serial tumor studies.

CSC can also be identified by efflux of incorporated Hoechst dyes via multidrug resistance (MDR) and ATP-binding cassette (ABC) Transporters.

Another approach which has also been used for identification of cell subsets enriched with CSCs *in vitro* is sphere-forming assays.

Many normal stem cells such as hematopoietics or stem cells from tissues are capable, under special culture conditions, to form three-dimensional spheres, which can differentiate into multiple cell types. As with normal stem cells, the CSCs isolated from brain or prostate tumors also have the ability to form anchorage-independent spheres.

Heterogeneity (CSC Markers)

Data over recent years have indicated the existence of CSCs in various solid tumors. For isolating CSCs from solid and hematological tumors, markers specific for normal stem cells of the same organ are commonly used.

Nevertheless, a number of cell surface markers have proved useful for isolation of subsets enriched for CSC including CD133 (also known as PROM1), CD44, CD24, EpCAM (epithelial cell adhesion molecule, also known as epithelial specific antigen, ESA), THY1 and ATP-binding cassette B5 (ABCB5).

CD133 (prominin 1) is a five-transmembrane domain glycoprotein expressed on $CD34^+$ stem and progenitor cells, in endothelial precursors and fetal neural stem cells. It has been detected using its glycosylated epitope know as AC133.

EpCAM (epithelial cell adhesion molecule, ESA, TROP1) is hemophilic Ca^{2+}-independent cell adhesion molecule expressed on the basolateral surface of most epithelial cells.

CD90 (THY1) is a glycosylphosphatidylinositol glycoprotein anchored in the plasma membrane and involved in signal transduction. It may also mediate adhesion between thymocytes and thymic stroma.

CD44 (PGP1) is an adhesion molecule that has pleiotropic roles in cell signaling, migration and homing. It has multiple isoforms, including CD44H, which exhibits high affinity for hyaluronate, and CD44V which has metastatic properties.

CD24 (HSA) is a glycosylated glycosylphosphatidylinositol-anchored adhesion molecule, which has co-stimulatory role in B and T cells.

ALDH is a ubiquitous aldehyde dehydrogenase family of enzymes, which catalyzes the oxidation of aromatic aldehydes to carboxyl acids. For instance, it has role in conversion of retinol to retinoic acid, which is essential for survival.

The first solid malignancy from which CSCs were isolated and identified was breast cancer. Therefore, these CSCs are the most intensely studied. Breast CSCs have been enriched in $CD44^{+}CD24^{-/low}$, SP, $ALDH^{+}$ subpopulations. However, recent evidence indicates that breast CSCs are very phenotypically diverse, and there is evidence that not only CSC marker expression in breast cancer cells is heterogeneous but also there exist many subsets of breast CSC.

Last studies provide further support to this point. Both $CD44^{+}CD24^{-}$ and $CD44^{+}CD24^{+}$ cell populations are tumor initiating cells; however, CSC are most highly enriched using the marker profile $CD44^{+}CD49f^{hi}CD133/2^{hi}$.

CSCs have been reported in many brain tumors. Stem-like tumor cells have been identified using cell surface markers including CD133, SSEA-1 (stage-specific embryonic antigen-1), EGFR and CD44.

However, there is uncertainty about the use of CD133 for identification of brain tumor stem-like cells because tumorigenic cells are found in both $CD133^{+}$ and $CD133^{-}$ cells in some gliomas, and some $CD133^{+}$ brain tumor cells may not possess tumor-initiating capacity.

Similarly, CSCs have also been reported in human colon cancer. For their identification, cell surface markers such as CD133, CD44 and ABCB5, or functional analysis including clonal analysis or Aldefluor assay were used. Using CD133 as a positive marker for colon CSCs has generated conflicting results. Nevertheless, recent studies indicated that the AC133 epitope, but not the CD133 protein, is specifically expressed in colon CSCs and its expression is lost upon differentiation. In addition, using $CD44^{+}$ colon cancer cells and additional sub-fractionation of $CD44^{+}EpCAM^{+}$ cell population with CD166 enhance the success of tumor engraftments.

Multiple CSCs have been reported in prostate, lung and many other organs, including liver, pancreas, kidney or ovary. In prostate cancer, the tumor-initiating cells have been identified in $CD44^{+}$ cell subset as $CD44^{+}$á2â1^{+}, TRA-1-$60^{+}CD151^{+}CD166^{+}$ or $ALDH^{+}$ cell populations. Putative markers for lung CSCs have been reported, including $CD133^{+}$, $ALDH^{+}$, $CD44^{+}$ and oncofetal protein $5T4^{+}$.

Metastatic Cancer Stem Cells

Metastasis is the major cause of tumor lethality in patients. However, not every cell in the tumor has the ability to metastasize. This potential depends on factors that determine growth, angiogenesis, invasion and other basic processes of tumor cells. In the many epithelial tumors, the epithelial-mesenchymal transition (EMT) is considered as a crucial events in the metastatic process.

EMT and the reverse transition from mesenchymal to an epithelial phenotype (MET) are involved in embryonic development, which involves disruption of epithelial cell homeostasis and the acquisition of a migratory mesenchymal phenotype.

The EMT appears to be controlled by canonical pathways such as WNT and transforming growth factor â pathway. The important feature of EMT is the loss of membrane E-cadherin in adherent junctions, where the â-catenin may play a significant role.

Translocation of â-catenin from adherent junctions to the nucleus may lead to a loss of E-cadherin, and subsequently to EMT. There is evidence that nuclear â-catenin can directly transcriptionally activate EMT-associated target genes, such as the E-cadherin gene repressor SLUG (also known as SNAI2).

Recent data have supported the concept, that tumor cells undergoing an EMT could be precursors for metastatic cancer cells, or even metastatic CSCs. In the invasive edge of pancreatic carcinoma a subset of $CD133^+CXCR4^+$ (receptor for CXCL12 chemokine also known as a SDF1 ligand) cells has been defined.

These cells exhibited significantly stronger migratory activity than their counterpart $CD133^+CXCR4^-$ cells, but both cell subsets showed similar tumor development capacity.

Moreover, inhibition of the CXCR4 receptor led to the reduced metastatic potential without altering tumorigenic capacity.

On the other hand, in the breast cancer $CD44^+CD24^{-/low}$ cells are detectable in metastatic pleural effusions. By contrast, an increased number of $CD24^+$ cells have been identified in distant metastases in patients with breast cancer.

Although, there are only few data on mechanisms mediating metastasis in breast cancer, it is possible that $CD44^+CD24^{-/low}$ cells initially metastasize and in the new site they change their phenotype and undergo limited differentiation.

These findings led to new dynamic two-phase expression pattern concept based on the existence of two forms of cancer stem cells - stationary cancer stem cells (SCS) and mobile cancer stem cells (MCS). SCS are embedded in tissue and persist in differentiated areas throughout all tumor progression.

The term MCS describes cells that are located at the tumor-host interface. There is an evidence that these cells are derived from SCS through the acquisition of transient EMT

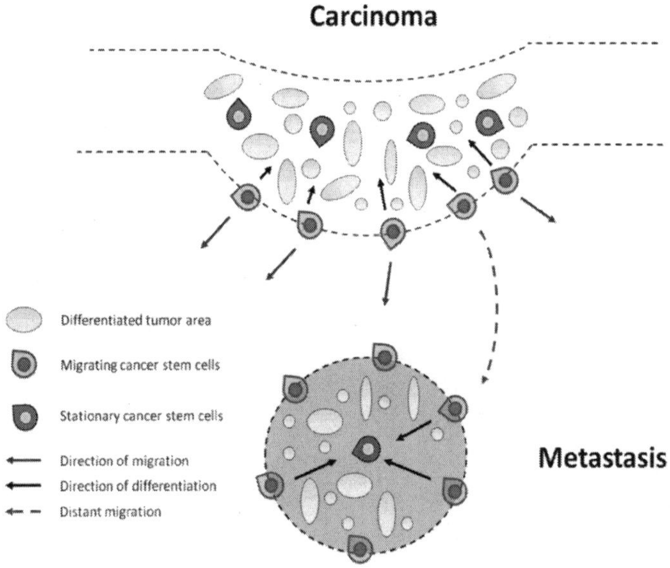

Figure: *The concept of migrating cancer stem cells (MSC). Stationary cancer stem cells are embedded in begin carcinomas and these cells are detectable in the differentiated central area of a tumor. The important step toward malignancy is the induction of epithelial mesenchymal transition (EMT) in the stationary cancer stem cells (SCS), which become mobile or migrating cancer stem cells. MCS cells divide asymmetrically. One daughter cell starts proliferation and differentiation. The remaining MCS migrates a short distance before undergoing new asymmetric division, or starts dissemination through blood vessela or lymphatic vessels and produces a metastasis.*

Implications for Cancer Treatment

The existence of CSCs has several implications in terms of future cancer treatment and therapies. These include disease identification, selective drug targets, prevention of metastasis, and development of new intervention strategies.

Normal somatic stem cells are naturally resistant to chemotherapeutic agents. They produce various pumps (such as MDR) that pump out drugs and DNA repair proteins and they also have a slow rate of cell turnover (chemotherapeutic agents naturally target rapidly replicating cells). CSCs that develop from normal stem cells may also produce these proteins, which could increase their resistance towards chemotherapeutic agents. The surviving CSCs then repopulate the tumor, causing a relapse.

By selectively targeting CSCs, it would be possible to treat patients with aggressive, non-resectable tumors, as well as preventing patients from metastasizing and relapsing. The hypothesis suggests that upon CSC elimination, cancer could regress due to differentiation and/or cell death. What fraction of tumor cells are CSCs and therefore need to be eliminated is not clear yet.

A number of studies have investigated the possibility of identifying specific markers that may distinguish CSCs from the bulk of the tumor (as well as from normal stem cells). Proteomic and genomic signatures of tumors are also being investigated.

In 2009, scientists identified one compound, Salinomycin, that selectively reduces the proportion of breast CSCs in mice by more than 100-fold relative to Paclitaxel, a commonly used chemotherapeutic agent.

The cell surface receptor interleukin-3 receptor-alpha (CD123) was shown to be overexpressed on CD34+CD38- leukemic stem cells (LSCs) in acute myelogenous leukemia (AML) but not on normal CD34+CD38- bone marrow cells. Jin et al., then demonstrated that treating AML-engrafted NOD/SCID mice with a CD123-specific monoclonal antibody impaired LSCs homing to the bone marrow and reduced overall AML cell repopulation including the proportion of LSCs in secondary mouse recipients.

Pathways

The design of new drugs for the treatment of CSCs will likely require an understanding of the cellular mechanisms that regulate cell proliferation. The first advances in this area were made with hematopoietic stem cells (HSCs) and their transformed counterparts in leukemia, the disease for which the origin of CSCs is best understood. It is now becoming increasingly clear that stem cells of many organs share the same cellular pathways as leukemia-derived HSCs.

Additionally, a normal stem cell may be transformed into a cancer stem cell through disregulation of the proliferation and differentiation pathways controlling it or by inducing oncoprotein activity.

BMI-1

The Polycomb group transcriptional repressor Bmi-1 was discovered as a common oncogene activated in lymphoma and later shown to specifically regulate HSCs. The role of Bmi-1 has also been illustrated in neural stem cells. The pathway appears to be active in CSCs of pediatric brain tumors.

Notch

The Notch pathway has been known to developmental biologists for decades. Its role in control of stem cell proliferation has now been demonstrated for several cell types including hematopoietic, neural and mammary stem cells. Components of the Notch pathway have been proposed to act as oncogenes in mammary and other tumors.

A particular branch of the Notch signaling pathway that involves the transcription factor Hes3 has been shown to regulate a number of cultured cells with cancer stem cell characteristics obtained from glioblastoma patients.

Sonic Hedgehog and Wnt

These developmental pathways are also strongly implicated as stem cell regulators. Both Sonic hedgehog (SHH) and Wnt pathways are commonly hyperactivated in tumors and are required to sustain tumor growth. However, the Gli transcription factors that are regulated by SHH take their name from gliomas, where they are commonly expressed at high levels.

A degree of crosstalk exists between the two pathways and their activation commonly goes hand-in-hand. This is a trend rather than a rule. For instance, in colon cancer hedgehog signalling appears to antagonise Wnt.

Sonic hedgehog blockers are available, such as cyclopamine. There is also a new water soluble cyclopamine that may be more effective in cancer treatment. There is also DMAPT, a water soluble derivative of parthenolide (induces oxidative stress, inhibits NF-êB signaling) for AML (leukemia), and possibly myeloma and prostate cancer.

A clinical trial of DMAPT is to start in England in late 2007 or 2008. Finally, the enzyme telomerase may qualify as a study subject in CSC physiology. GRN163L (Imetelstat) was recently started in trials to target myeloma stem cells. If it is possible to eliminate the cancer stem cell, then a potential cure may be achieved if there are no more CSCs to repopulate a cancer.

Breast Cancer

Breast cancer is a type of cancer originating from breast tissue, most commonly from the inner lining of milk ducts or the lobules that supply the ducts with milk.

Cancers originating from ducts are known as ductal carcinomas, while those originating from lobules are known as lobular carcinomas. Breast cancer occurs in humans and other mammals. While the overwhelming majority of human cases are in women, breast cancer can also occur in men.

The balance of benefits versus harms of breast cancer screening is controversial. The characteristics of the cancer determine the treatment, which may include surgery, medications (hormonal therapy and chemotherapy), radiation and/or immunotherapy.

Surgery provides the single largest benefit, and to increase the likelihood of remission (no further sign of the cancer), several chemotherapy regimens are commonly given in addition. Radiation is used after breast-conserving surgery and substantially improves local relapse rates and in many circumstances also overall survival.

A considerable part of the current knowledge on breast carcinomas is based on in vivo and in vitro studies performed with cell lines derived from breast cancers.

These provide an unlimited source of homogenous self-replicating material, free of contaminating stromal cells, and often easily cultured in simple standard media. The first breast cancer cell line described, BT-20, was established in 1958. Since then, and despite sustained work in this area, the number of permanent lines obtained has been strikingly low (about 100).

Indeed, attempts to culture breast cancer cell lines from primary tumors have been largely unsuccessful. This poor efficiency was often due to technical difficulties associated with the extraction of viable tumor cells from their surrounding stroma. Most of the available breast cancer cell lines issued from metastatic tumors, mainly from pleural effusions.

Effusions provided generally large numbers of dissociated, viable tumor cells with little or no contamination by fibroblasts and other tumor stroma cells. Many of the currently used BCC lines were established in the late 1970s. A very few of them, namely MCF-7, T-47D, and MDA-MB-231, account for more than two-thirds of all abstracts reporting studies on mentioned breast cancer cell lines, as concluded from a Medline-based survey.

Dendrimers as Drug Carriers

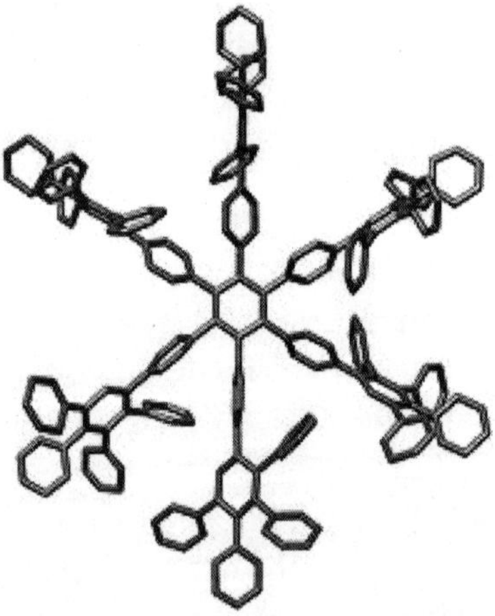

Figure: *Crystal structure of a first-generation polyphenylene dendrimer reported by Müllen et al.*

Dendrimers are repetitively branched molecules. The name comes from the Greek word äÝíäñïí (dendron), which translates to "tree". Synonymous terms for dendrimer include arborols and cascade molecules. However, dendrimer is currently the internationally accepted term. A dendrimer is typically symmetric around the core, and often adopts a spherical three-dimensional morphology.

The word dendron is also encountered frequently. A dendron usually contains a single chemically addressable group called the focal point. The difference between dendrons and dendrimers is illustrated in figure one, but the terms are typically encountered interchangeably.

The first dendrimers were made by divergent synthesis approaches by Fritz Vögtle in 1978, R.G. Denkewalter at Allied Corporation in 1981, Donald Tomalia at Dow Chemical in 1983 and in 1985, and by George Newkome in 1985. In 1990 a convergent synthetic approach was introduced by Jean Fréchet. Dendrimer popularity then greatly increased, resulting in more than 5,000 scientific papers and patents by the year 2005.

Properties

Dendritic molecules are characterized by structural perfection. Dendrimers and dendrons are monodisperse and usually highly symmetric, spherical compounds.

The field of dendritic molecules can be roughly divided into low-molecular weight and high-molecular weight species. The first category includes dendrimers and dendrons, and the latter includes dendronized polymers, hyperbranched polymers, and the polymer brush.

The properties of dendrimers are dominated by the functional groups on the molecular surface, however, there are examples of dendrimers with internal functionality. Dendritic encapsulation of functional molecules allows for the isolation of the active site, a structure that mimics that of active sites in biomaterials.

Also, it is possible to make dendrimers water soluble, unlike most polymers, by functionalizing their outer shell with charged species or other hydrophilic groups. Other controllable properties of dendrimers include toxicity, crystallinity, tecto-dendrimer formation, and chirality.

Dendrimers are also classified by generation, which refers to the number of repeated branching cycles that are performed during its synthesis. For example if a dendrimer is made by convergent synthesis, and the branching reactions are performed onto the core molecule three times, the resulting dendrimer is considered a third generation dendrimer.

Each successive generation results in a dendrimer roughly twice the molecular weight of the previous generation.

Higher generation dendrimers also have more exposed functional groups on the surface, which can later be used to customize the dendrimer for a given application.

Synthesis

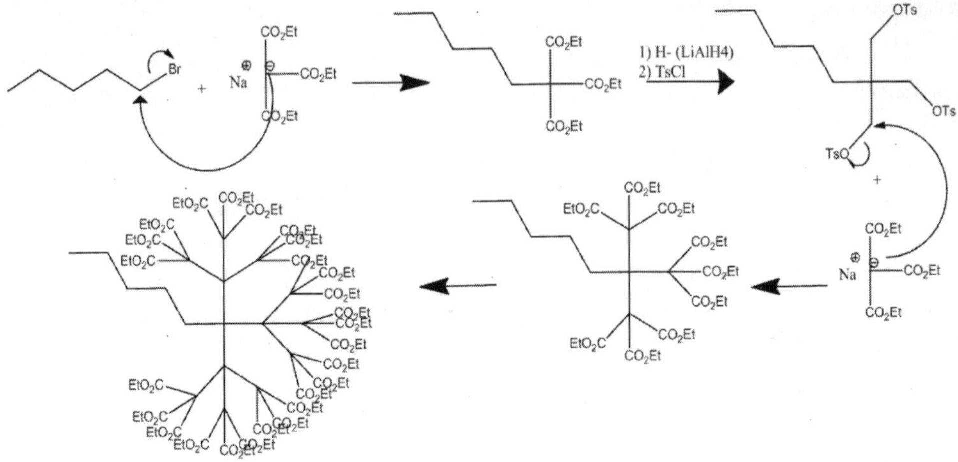

Figure: *Synthesis to second generation arborol*

One of the very first dendrimers, the Newkome dendrimer, was synthesized in 1985. This macromolecule is also commonly known by the name arborol. The synthesis is started by nucleophilic substitution of 1-bromopentane by *triethyl sodiomethanetricarboxylate* in dimethylformamide and benzene. The ester groups were then reduced by lithium aluminium hydride to a triol in a deprotection step.

Activation of the chain ends was achieved by converting the alcohol groups to tosylate groups with tosyl chloride and pyridine. The tosyl group then served as leaving groups in another reaction with the tricarboxylate, forming generation two. Further repetition of the two steps leads to higher generations of arborol.

Poly(amidoamine), or PAMAM, is perhaps the most well known dendrimer. The core of PAMAM is a diamine (commonly ethylenediamine), which is reacted with methyl acrylate, and then another ethylenediamine to make the generation-0 (G-0) PAMAM. Successive reactions create higher generations, which tend to have different properties.

Lower generations can be thought of as flexible molecules with no appreciable inner regions, while medium sized (G-3 or G-4) do have internal space that is essentially separated from the outer shell of the dendrimer. Very large (G-7 and greater) dendrimers can be thought of more like solid particles with very dense surfaces due to the structure of their outer shell. The functional group on the surface of PAMAM dendrimers is ideal for click chemistry, which gives rise to many potential applications.

Dendrimers can be considered to have three major portions: a core, an inner shell, and an outer shell. Ideally, a dendrimer can be synthesized to have different functionality in each of these portions to control properties such as solubility, thermal stability, and attachment of compounds for particular applications. Synthetic processes can also precisely control the size and number of branches on the dendrimer.

There are two defined methods of dendrimer synthesis, divergent synthesis and convergent synthesis.

However, because the actual reactions consist of many steps needed to protect the active site, it is difficult to synthesize dendrimers using either method. This makes dendrimers hard to make and very expensive to purchase.

At this time, there are only a few companies that sell dendrimers; Polymer Factory Sweden AB commercializes biocompatible bis-MPA dendrimers and Dendritech is the only kilogram-scale producers of PAMAM dendrimers.

NanoSynthons, LLC from Mount Pleasant, Michigan, USA produces PAMAM dendrimers and other proprietary dendrimers.

Weihai CY Dendrimer Technology Co. Ltd. is the 1st enterprise in Asia with creative & high technologies to research, develop & produce dendrimers & hyperbranched polymers.

The Dendrimer material has been well used for scientific research in many colleges and universities such as Peking University, Tsinghua University, Chinese Academy of Sciences, and began to be industrialized in China now.

Divergent Methods

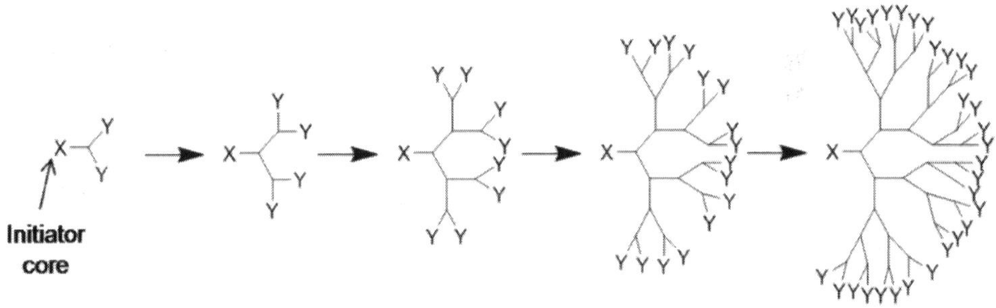

Figure: *Schematic of divergent synthesis of dendrimers*

The dendrimer is assembled from a multifunctional core, which is extended outward by a series of reactions, commonly a Michael reaction. Each step of the reaction must be driven to full completion to prevent mistakes in the dendrimer, which can cause trailing generations (some branches are shorter than the others).

Such impurities can impact the functionality and symmetry of the dendrimer, but are extremely difficult to purify out because the relative size difference between perfect and imperfect dendrimers is very small.

Convergent Methods

Dendrimers are built from small molecules that end up at the surface of the sphere, and reactions proceed inward building inward and are eventually attached to a core. This method makes it much easier to remove impurities and shorter branches along the way, so that the final dendrimer is more monodisperse.

However dendrimers made this way are not as large as those made by divergent methods because crowding due to steric effects along the core is limiting.

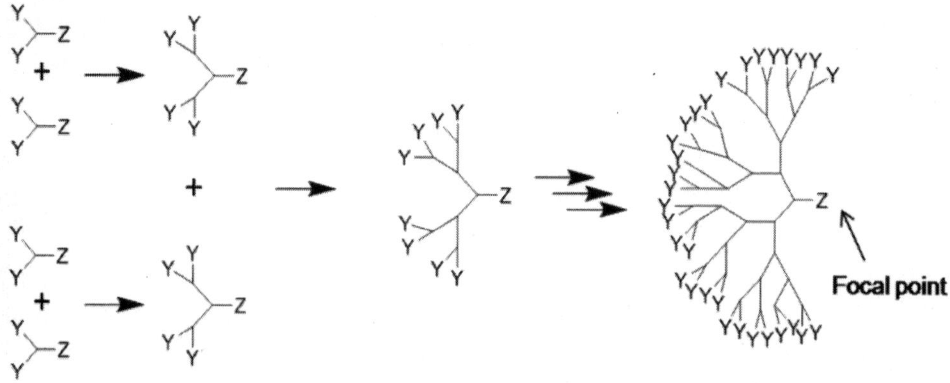

Figure: *Schematic of convergent synthesis of dendrimers*

Click Chemistry

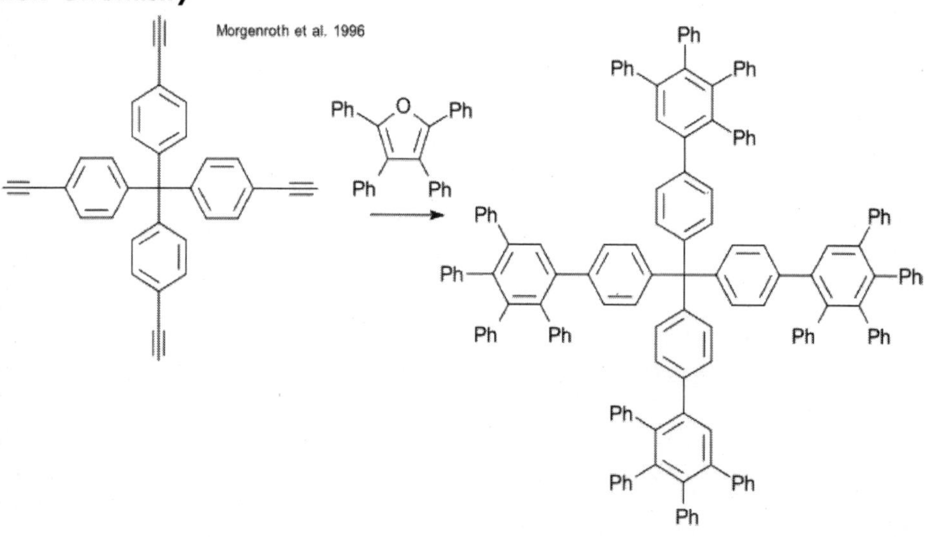

Figure: *Dendrimer DA reaction Mullen 1996*

Dendrimers have been prepared via click chemistry, employing Diels-Alder reactions, thiol-yne reactions and azide-alkyne reactions.

An example is the synthesis of certain polyphenylene dendrimers can be seen in figure earlier.

There are ample avenues that can be opened by exploring this chemistry in dendrimer synthesis.

Applications

Applications of dendrimers typically involves conjugating other chemical species to the dendrimer surface that can function as detecting agents (such as a dye molecule), affinity ligands, targeting components, radioligands, imaging agents, or pharmaceutically active compounds.

Dendrimers have very strong potential for these applications because their structure can lead to multivalent systems.

In other words, one dendrimer molecule has hundreds of possible sites to couple to an active species. Researchers aimed to utilise the hydrophobic environments of the dendritic media to conduct photochemical reactions that generate the products that are synthetically challenged. Carboxylic acid and phenol terminated water soluble dendrimers were synthesized to establish their utility in drug delivery as well as conducting chemical reactions in their interiors.

This might allow researchers to attach both targeting molecules and drug molecules to the same dendrimer, which could reduce negative side effects of medications on healthy cells.

Dendrimers can also be used as a solubilizing agent. Since their introduction in the mid-1980s, this novel class of dendrimer architecture has been a prime candidate for host-guest chemistry. Dendrimers with hydrophobic core and hydrophilic periphery have shown to exhibit micelle-like behaviour and have container properties in solution.

The use of dendrimers as unimolecular micelles was proposed by Newkome in 1985. This analogy highlighted the utility of dendrimers as solubilizing agents. The majority of drugs available in pharmaceutical industry are hydrophobic in nature and this property in particular creates major formulation problems.

This drawback of drugs can be ameliorated by dendrimeric scaffolding, which can be used to encapsulate as well as to solubilize the drugs because of the capability of such scaffolds to participate in extensive hydrogen bonding with water.

Dendrimer labs throughout the planet are persistently trying to manipulate dendrimer's solubilizing trait, in their way to explore dendrimer as drug delivery and target specific carrier.

Drug Delivery

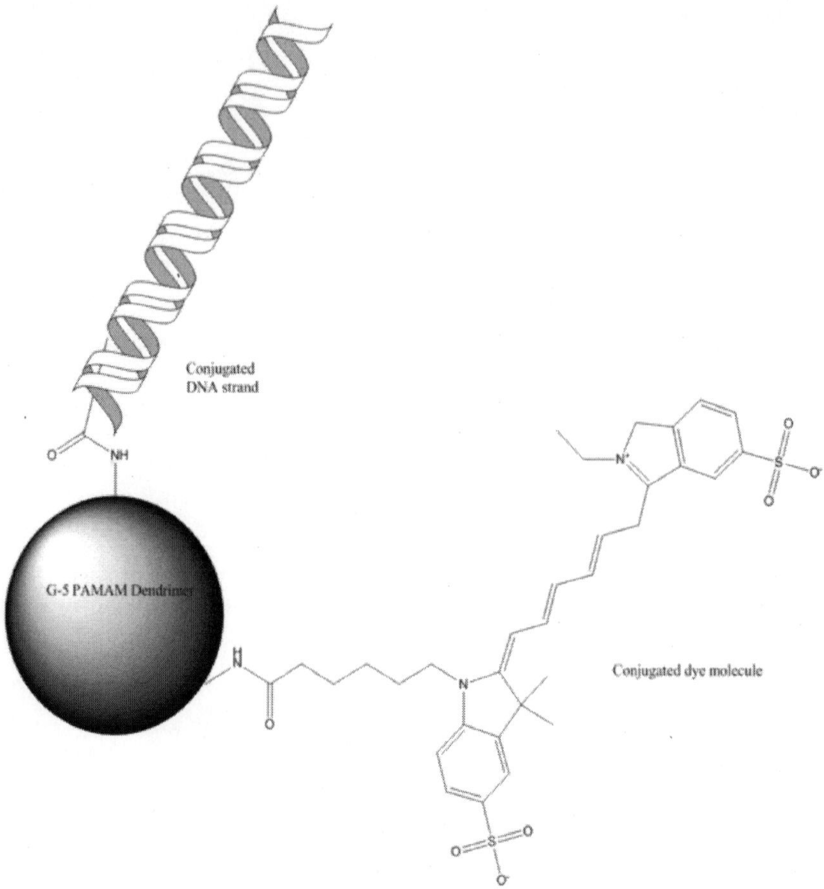

Figure: *Schematic of a G-5 PAMAM dendrimer conjugated to both a dye molecule and a strand of DNA.*

Approaches for delivering unaltered natural products using polymeric carriers is of widespread interest, dendrimers have been explored for the encapsulation of hydrophobic compounds and for the delivery of anticancer drugs. The physical characteristics of dendrimers, including their monodispersity, water solubility, encapsulation ability, and large number of functionalizable peripheral groups, make these macromolecules appropriate candidates for evaluation as drug delivery vehicles.

There are three methods for using dendrimers in drug delivery: first, the drug is covalently attached to the periphery of the dendrimer to form dendrimer prodrugs, second the drug is coordinated to the outer functional groups via ionic interactions, or third the dendrimer acts as a unimolecular micelle by encapsulating a pharmaceutical through the formation of a dendrimer-drug supramolecular assembly.

The use of dendrimers as drug carriers by encapsulating hydrophobic drugs is a potential method for delivering highly active pharmaceutical compounds that may not be in clinical use due to their limited water solubility and resulting suboptimal pharmacokinetics. Dendrimers have been widely explored for controlled delivery of antiretroviral bioactives.

The inherent antiretroviral activity of dendrimers enhances their efficacy as carriers for antiretroviral drugs The dendrimer enhances both the uptake and retention of compounds within cancer cells, a finding that was not anticipated at the onset of studies.

The encapsulation increases with dendrimer generation and this method may be useful to entrap drugs with a relatively high therapeutic dose. Studies based on this dendritic polymer also open up new avenues of research into the further development of drug-dendrimer complexes specific for a cancer and/or targeted organ system. These encouraging results provide further impetus to design, synthesize, and evaluate dendritic polymers for use in basic drug delivery studies and eventually in the clinic.

Gene Delivery

The ability to deliver pieces of DNA to the required parts of a cell includes many challenges. Current research is being performed to find ways to use dendrimers to traffic genes into cells without damaging or deactivating the DNA. To maintain the activity of DNA during dehydration, the dendrimer/DNA complexes were encapsulated in a water soluble polymer, and then deposited on or sandwiched in functional polymer films with a fast degradation rate to mediate gene transfection. Based on this method, PAMAM dendrimer/DNA complexes were used to encapsulate functional biodegradable polymer films for substratemediated gene delivery. Research has shown that the fast degrading functional polymer has great potential for localized transfection.

Sensors

Scientists have also studied dendrimers for use in sensor technologies. Studied systems include proton or pH sensors using poly(propylene imine), cadmium-sulfide/polypropylenimine tetrahexacontaamine dendrimer composites to detect fluorescence signal quenching, and poly(propylenamine) first and second generation dendrimers for metal cation photodetection amongst others. Research in this field is vast and ongoing due to the potential for multiple detection and binding sites in dendritic structures.

Blood Substitution

Dendrimers are also being investigated for use as blood substitutes. Their steric bulk surrounding a heme-mimetic centre significantly slows degradation compared to free heme, and prevents the cytotoxicity exhibited by free heme.

Nanoparticles

Dendrimers also are used in the synthesis of monodisperse metallic nanoparticles. Poly(amidoamide), or PAMAM, dendrimers are utilised for their tertiary amine groups at the branching points within the dendrimer. Metal ions are introduced to an aqueous dendrimer solution and the metal ions form a complex with the lone pair of electrons present at the tertiary amines. After complexion, the ions are reduced to their zerovalent states to form a nanoparticle that is encapsulated within the dendrimer. These nanoparticles range in width from 1.5 to 10 nanometres and are aptly called Dendrimer-Encapsulated Nanoparticles.

Microencapsulation:- poly(amido)amine (PAMAM) has been utilised to encapsulate self healing agent for smart coating by microencapsulation route. The idea has the potential to fabricate microcapsules which could provide better anticorrosive properties to polymeric coatings.

Nanotechnology in Drug Administration and Food

Nanotechnology involves the engineering of functional systems at the molecular scale. Such systems are characterized by unique physical, optical and electronic features that are attractive for disciplines ranging from materials science to biomedicine. One of the most active research areas of nanotechnology is nanomedicine, which applies nanotechnology to highly specific medical interventions for the prevention, diagnosis and treatment of diseases. The surge in nanomedicine research during the past few decades is now translating into considerable commercialization efforts around the globe, with many products on the market and a growing number in the pipeline. Currently, nanomedicine is dominated by drug delivery systems, accounting for more than 75% of total sales.

Nanomaterials fall into a size range similar to proteins and other macromolecular structures found inside living cells. As such, nanomaterials are poised to take advantage of existing cellular machinery to facilitate the delivery of drugs. Nanoparticles (NPs) containing encapsulated, dispersed, absorbed or conjugated drugs have unique characteristics that can lead to enhanced performance in a variety of dosage forms. When formulated correctly, drug particles are resistant to settling and can have higher saturation solubility, rapid dissolution and enhanced adhesion to biological surfaces, thereby providing rapid onset of therapeutic action and improved bioavailability. In addition, the vast majority of molecules in a nanostructure reside at the particle surface, which maximizes the loading and delivery of cargos, such as therapeutic drugs, proteins and polynucleotides, to targeted cells and tissues. Highly efficient drug delivery, based on nanomaterials, could potentially reduce the drug dose needed to achieve therapeutic benefit, which, in turn, would lower the cost and/or reduce the side effects associated with particular drugs. Furthermore, NP size and surface characteristics can be easily manipulated to achieve both passive and active drug targeting. Site-specific targeting can be achieved by attaching targeting ligands, such as antibodies or aptamers, to the surface of particles, or by using guidance in the form of magnetic NPs. NPs can also control and sustain release of a drug during transport to, or at, the site of localization, altering drug distribution and subsequent clearance of the drug in order to improve therapeutic efficacy and reduce side effects.

Nanotechnology could be strategically implemented in new developing drug delivery systems that can expand drug markets. Such a plan would be applied to drugs selected for full-scale development based on their safety and efficacy data, but which fail to reach clinical development because of poor biopharmacological properties, for example, poor solubility or poor permeability across the intestinal epithelium, situations that translate into poor bioavailability and undesirable pharmacokinetic properties. The new drug delivery methods are expected to enable pharmaceutical companies to reformulate existing drugs on the market, thereby extending the lifetime of products and enhancing the performance of drugs by increasing effectiveness, safety and patient adherence, and ultimately reducing healthcare costs.

Applications of Nanosystems in Cosmetics

Nanoscientists design, develop and supervise the production of materials, devices and systems of unique molecular or macromolecular composition, applying principles of nanoscale physics and electrical, chemical and biological engineering.

Nanoscientists study structures of 100 nanometres (nm) or less. Since a nanometre only measures one billionth of a metre, it's difficult to imagine anything smaller. Nanotechnology is expected to become one of the most strategic and dominant technologies in the next 10 to 20 years. Nanotechnology will have an almost endless string of applications in biotechnology, biology and biomedicine.

Nanotechnology has had several commercial applications in advanced laser technology, hard coatings, photography, pharmaceuticals, printing, chemical-mechanical polishing and cosmetics. Soon there will be lighter cars using nanoparticle-reinforced polymers, insulin that can be taken orally, artificial joints made from nanoparticulate materials and low-kilojoule foods with nanoparticulate taste enhancers.

Nanoscientists create designs or prototypes for nanosystem applications, such as biomedical delivery systems and atomic force microscopes.

They design or engineer nanodevices etc. using three-dimensional computer-aided design (CAD) software. They coordinate or supervise the work of suppliers or vendors in the designing, building or testing of nanosystem devices, such as lenses or probes.

They design or conduct tests on new nanotechnology products, processes, or systems and engineer production processes for specific nanotechnology applications, such as electroplating, nanofabrication, or epoxy. They conduct research related to a range of nanotechnology topics, such as packaging, heat-transfer, fluorescence detection, nanoparticle dispersion, hybrid systems, liquid systems, nanocomposites, nanofabrication, optoelectronics and nanolithography.

Daily tasks include lab work and analysis, meetings with students and internal meetings, computer work, articles to write, numerous discussions about research findings and much reading to keep up with current information.

In areas as diverse as designing medical diagnostic devices to building better batteries, from creating cosmetics to enhancing energy efficient windows, from auto and plane manufacturing to researching the nature of matter itself, knowledge of nanoscale science and technology will be increasingly important during upcoming years and decades.

As nanoscale science and technology come to have increasing impacts on many aspects of our daily lives, the opportunities for careers in these fields are expanding rapidly.

Targeting Cancer with Engineered Viruses and Immune Cells

Targeting is the future of cancer treatment. The fact that treatments can be targeted to cancer cells with even a moderate level of precision is the important factor, not the nature of the mechanism used to destroy those cells. Targeting means few or no side effects, which in turn means that more potent attacks can be conducted on cancerous cells: no more will cancer therapy be a matter of finely juggling whole-body chemical infusions to kill as much of the cancer as possible without killing or crippling the patient.

Many different cell killing mechanisms have been demonstrated in the laboratory in conjunction with forms of targeting - there are any number of ways to sabotage a cell to the point at which it self-destructs, such as by dumping in enough heat, or radiation, or toxic molecules, or carefully selected sabotaging proteins that throw a spanner into cellular machinery. When nanoparticles are used as the delivery platform, researchers can even load them up with old-school chemotherapy drugs: minute amounts attached to each nanoparticle, but enormous doses for each individual cancer cell in comparison to what would be received in the standard form of presently widespread chemotherapy.

Here are two examples of other methods of targeting cancer presently under development: a gene therapy delivered by engineered virus, and altered immune cells that are programmed to attack and destroy cells with cancerous surface markers.

Targeted viral therapy destroys breast cancer stem cells in preclinical experiments

A promising new treatment for breast cancer [has] been shown in cell culture and in animal models to selectively kill cancer stem cells at the original tumor site and in distant metastases with no toxic effects on healthy cells, including normal stem cells. Melanoma differentiation associated gene-7 (mda-7), also known as interleukin (IL)-24, has been shown to directly impact two forms of cell suicide known as apoptosis and toxic autophagy, regulate the development of new blood vessels and also play a role in promoting cancer cell destruction by the immune system. In the present study, the researchers used a recombinant adenovirus vector, an engineered virus with modified genetic material, known as Ad.mda-7 to deliver the mda-7/IL-24 gene with its encoded protein directly to the tumor.

Since discovering the mda-7/IL-24 gene, Fisher and his team have worked to develop better ways to deliver it to cancer cells, including two cancer "terminator" viruses known as Ad.5-CTV and Ad.5/3-CTV. Cancer terminator viruses are unique because they are designed to replicate only within cancer cells while delivering immune-modulating and toxic genes such as MDA-7/IL-24. Coupled with a novel stealth delivery technique known as ultrasound-targeted microbubble destruction (UTMD), researchers can now systemically deliver viruses and therapeutic genes and proteins directly to tumors and their surrounding tissue (microenvironment) at both primary and metastatic tumor sites. UTMD uses microscopic, gas-filled bubbles that can be paired with viral therapies, therapeutic genes and proteins, and imaging agents and can then be released in a site and target-specific manner via ultrasound. Fisher and his colleagues are pioneering this approach and have already reported success in experiments utilising UTMD technology and mda-7/IL-24 gene therapy in prostate and colorectal cancer models.

Modified Immune Cells Seek and Destroy Melanoma

All cells express a complex known as the proteasome, which acts as the garbage disposal for the cell. There are two types of proteasomes: constitutive proteasomes (cPs), which are found in normal tissues, and immunoproteasomes (iPs), which are found in stressed or damaged cells. In a damaged cell, the iP generates protein fragments that are displayed on the surface of the distressed cells, triggering recognition by dendritic cells and subsequent destruction by the immune system.

Most cancers, including melanoma, exclusively express cPs, making it impossible for them to express the protein fragments that are recognised by the immune system. To make it easier for the immune system to find cancer cells, [researchers] engineered a specific type of immune cell, known as a dendritic cell, that recognises protein fragments of cancer specific antigens made by cPs. The engineered dendritic cells were then injected into patients that were in remission from melanoma.

The trial consisted of 4 patients that were vaccinated with regular dendritic cells, 3 patients that received cells that underwent a control treatment, and 5 patients that received dendritic cells that recognised cancer-made protein fragments. Vaccination with all three types of dendritic cells elicited an immune response, which peaked after 3-4 vaccinations with dendritic cells. Patients that received the specially modified dendritic cells had a longer lasting immune response and fewer circulating melanoma cells. Of the two patients that had active disease, treatment with modified dendritic cells resulted in a partial clinical response in one and a complete clinical response in the other.

A number of other forms of targeting are further ahead in the process of development than the two examples above: there is a lot of variety in the approaches used, and that is a very good thing. Competition drives progress. It is not unreasonable to think that cancer will be a controllable part of our biology twenty years from now, gone the way of tuberculosis to become something that can only seriously harm you if you do not have access to clinical medicine.

Bibliography

Altmann, J. & Gubrud, M.A.: *Discovering the Nanoscale*, IOS Press, Amsterdam, 2004.

Andrew, N.: *Foundations of Nanomechanics: from Solid-state Theory to Device Applications*, Berlin, New York, 2003.

Attebery, B.: *Nanoculture: Implications of the New Technoscience*, Intellect Books, Bristol, 2004.

Bainbridge, W.S. : *Societal Implications of Nanoscience and Nanotechnology*, Dordrecht, Kluwer, 2001.

Bast, R.C. Jr.: *Cancer Medicine*, Decker, Hamilton, 2000.

Berube, D.M.: *Discovering the Nanoscale*, IOS Press, Amsterdam, 2004.

Brosnan, J.: *The Primal Scream: A History of Science Fiction Film*, Orbit Books, London, 1991.

Carroll, J.S.: *Social Science Research Methods for Assessing Societal Implications of Nanotechnology*, Dordrecht, Kluwer, 2001.

Caviedes, C.: *El Nino in History: Storming Through the Ages*, University Press of Florida, Florida, 2001.

Credi, A.: *Molecular Devices and Machines: a Journey into the Nano World*, Wiley-VCH, Weinheim, 2003.

Crow, M. & Sarewitz, D.: *Societal Implications of Nanoscience and Nanotechnology*, Dordrecht: Kluwer, 2001.

Daniel Ratner: *Nanotechnology: a Gentle Introduction to the Next Big Idea*, Upper Saddle River, Prentice Hall, 2003.

Dobzhanshy, T.: *Genetics of the Evolutionary Process*, Columbia University Press, New York, 1970.

Drexler, K. Eric: *Engines of Creation: The Coming Era of Nanotechnology*, Anchor Books, New York, 1986.

Dyson, Tim: *India's Historical Demography: Studies in Famine, Disease, and Society*, Riverdale, Westwood, MA, 1989.

Eamonn Gearon: *The Sahara: A Cultural History*, Signal Books, UK, 2011.

Frank J. Owens: *Introduction to Nanotechnology*, Hoboken, J. Wiley, 2003.

Glimell, H.: *Dynamics of the Emerging Field of Nanoscience*, Dordrecht, Kluwer, 2001.

Glimell, H.: *Societal Implications of Nanoscience and Nanotechnology*, Kluwer, Dordrecht, 2001.

Haber, L. F.: *The Chemical Industry: 1900-1930, International Growth and Technological Change*, Oxford: Claredon Press, 1971.

Hessenbruch, A.: *Nanotechnology and the Negotiation of Novelty*, Amsterdam, IOS Press, 2004.

James Lewis: *Prospects in Nanotechnology: Toward Molecular Manufacturing,* John Wiley & Sons, New York, 1995.

Karn, Barbara: *Nanotechnology and the Environment: Applications and Implications,* American Chemical Society, Washington D.C., 2004.

Keblinski, P. : *Encyclopedia of Nanoscience and Nanotechnology*, American Scientific, Los Angeles, Calif, USA, 2004.

Liu W.K., Karpov E.G., Park H.S.: *Nano Mechanics and Materials,* Wiley, Weinheim, 2006.

Mayr, E.: *Change of Genetic Environment and Evolution*, Allen and Unwin, London, 1954.

Mody, C.C.M.: *How Probe Microscopists Became Nanotechnologists, Discovering the Nanoscale*, IOS Press, Amsterdam, 2004.

Norio Taniguchi: *Nanotechnology: Integrated Processing Systems for Ultra-precision and Ultra-fine Products,* Oxford University Press, New York, 1996.

Orth, R. N.: *Nanometer Scale Antibody Patterning for Directed cell Immobilization and Stimulation,* Cornell University, Ithaca, 2002.

Pressman, J.: *Nano Narrative: A Parable from Electronic Literature*, Intellect Books, Bristol, 2004.

Pressman, J.: *Nanoculture: Implications of the New Technoscience*, Intellect Books, Bristol, UK, 2004.

Ratner, Daniel: *Nanotechnology: a Gentle Introduction to the Next Big Idea*, Prentice Hall, Upper Saddle River, 2003.

Robison, W.L.: *Discovering the Nanoscale*, IOS Press, Amsterdam, 2004.

Sarewitz, D.: *Nanotechnology and Societal Transformation,* Dordrecht, Kluwer, 2001.

Stephen M. Goodnick: *Transport in Nanostructures,* Cambridge University Press, Cambridge, 1997.

Tenner, E.: *Nanotechnology and Unintended Consequences*, , Dordrecht, Kluwer, 2001.

Vasudev Bhatnagar: *Cell Science and Technology*, Campus Books, Delhi, 2009.

Vermaas, P.E.: *Discovering the Nanoscale,* IOS Press, Amsterdam, 2004.

Weil, V.: *Ethical Issues in Nanotechnology*, Dordrecht, Kluwer, 2001.

Index

❑❑❑